HOUSING THE PIG

GERRY BRENT

FARMING PRESS LIMITED
Wharfedale Road, Ipswich, Suffolk IP1 4LG

First published 1986

British Library Cataloguing in Publication Data

Brent, Gerry
Housing the pig.
1. Swine——Housing
I. Title
636.4′0831   SF396.3

ISBN 0-85236-157-2

ISBN 0 85236 157 2

Phototypeset by Galleon Photosetting, Ipswich
Printed in Great Britain by Redwood Burn Limited, Trowbridge, Wiltshire

Contents

List of Layouts

Acknowledgements

IN ADDITION to those who had the arduous task of producing the typed text of this book, my thanks go to Adrian Harrison for the production of the sketches and to *Pig Farming* magazine for permission to reproduce the photographs.

Also to Bruce Brockway, director of the Farm Buildings Information Centre at the NAC, Stoneleigh, whose special assistance is acknowledged in reading parts of the draft text and imparting comment. Also to various members of ADAS for help in providing guidance to allow a summary of legislation matters to be made.

Particular thanks must be made to a good friend, skilled pig manager and experienced builder of piggeries—Nick White, manager to Mr Roger Mercer in Staffordshire. There were times when the text was more like a book by Nick White than the author; this painstaking copy reading helped to make the text more comprehensible and clear and is acknowledged with many thanks.

Using this Book

MANY OF US learn the facts about pig buildings and equipment from hard experience. Some of the lessons are learnt from mistakes and so we tend to believe that we can avoid errors in the future.

The problem is that technology keeps changing, so an existing pig producer needs to know how to evaluate a new system or material by the time the next set of works are to be conducted.

Furthermore, the new producer requires a set of practical guidelines on how to go about developing or redeveloping a set of buildings.

Frequently buildings work less well because there was a basic failure to provide the correct size or number of pens for the classes of pigs kept. So planning and scheduling are given a comprehensive airing in this book after the general matters which influence the decision on pig developments, including a practical interpretation of the U.K. laws on planning and approvals.

The whole intention of the book is to provide a critical appraisal of all those matters which concern anyone planning to invest in buildings or pig equipment. It shows how to judge the components of a package-deal quotation, how to check on a contractor's specification, how to evaluate the materials proposed for use and how to decide which layout to use. Indeed later chapters include descriptions of building systems and layouts not just from a dimensional standpoint but also taking operation into account.

Unlike other books on buildings this will not lead a technician to research references. It will not be instantly usable to allow a building to be designed but it *will* allow the farmer to make an appraisal of proposed building work and to avoid some of the pitfalls—minor and major—into which many would otherwise stumble.

It is considered important that the author does *not* impose personal preferences upon the user of this book. Rather that the basic data is provided with comments and observations incorporated based upon actual experience.

This underlines the most important factors in the construction of pig buildings—that those who propose to construct pig accommodation work through the whole thought process, gathering the information which may best satisfy their particular circumstances and their own preferences. This book aims to provide guidelines on what information is required in order that such judgements may be made—including the vital operational considerations. The book may thus *influence* decisions—it cannot, and does not attempt to make arbitrary judgements upon a farmer's final choice of site or system.

Chapter 1

MODERN PIG HOUSING

THE DEVELOPMENT OF INTENSIVE PIG KEEPING

TRACING THE pattern of what is currently perceived as modern pig housing systems is not difficult because much of the development has occurred during the working lives of many people who are still involved in the pig industry. A genuine revolution has taken place, largely originating from the change to specialist pig enterprises from mixed farming. In some countries this began in the late 1940s and it was seen in the United Kingdom in the 1950s. It accelerated everywhere with great speed in the 1960s.

The practice of keeping pigs in larger 'intensive' or 'confinement' systems triggered off a need for higher capital investment. This in its turn stimulated the need for a higher level of productive efficiency to justify the expenditure and to service charges on borrowed capital. In addition, as unit size grew, the larger industrial type units increased the degree of international commodity trading in pigmeat and subjected pig producers to economic pressures outside their own control. This again emphasised the necessity for still greater productive efficiency.

All aspects of production have been affected by these phenomena and it would be incorrect to suggest that it is only housing systems which have been influenced by the change from the typical 30-sow herd in the 1950s to a current sow herd possibly ten times larger. Genetics and nutrition have been given close attention and there has been an increased understanding of the problem of disease as well as a proliferation of management aids. Together these have resulted in massive change. However, little specific genetic work has been undertaken to produce breeding stock suited to a particular housing system, save possibly the outdoor sow, or a robust pig suited to more rigorous conditions. Nutritional developments have also tended to respond to the effects of housing and genetic change rather than to initiate them. Operational management and stockmanship have to be developed if new production methods are to be successfully adopted. It is indeed the case that the operator holds the key to the achievement of successful production from any system.

The prime aim of the development of 'intensive' housing was greater output per man. For many years a significant body of opinion could not understand why a particular housing method yielded good results on one farm

1

but less satisfactory results on another. It is now better understood that the successful manager is the one who welds together a method of operation and feeding to suit the type of pig he has, with the housing systems available to him, given the particular health status of that population. Right at the outset therefore a conflict will exist between the initial need to increase output per man and the direct and critical effect that stockmanship has upon any pig enterprise.

The pursuit of increased labour productivity led to housing systems with a reduced reliance on manual inputs, particularly in manure removal and feed distribution. The adoption of slatted systems of housing became common, ruling against the use of bedding which had always been the buffer against environmental shortcomings in pig buildings up to this time. The removal of bedding brought about another significant change. More mechanical temperature control and wider use of fully insulated structures became essential, even if initially both were poorly conceived and reluctantly accepted. The removal of bedding also served to highlight such factors as good floor construction and the importance of appropriate stocking rates in these building systems.

In the wake of these financial and operational changes came the quest to keep more pigs in the same or, preferably, smaller area. Coupled with the removal of bedding, this gave rise to a new group of management problems, particularly vices such as tail biting and ear sucking. Counter-measures adopted to mitigate against them included reduction of stocking rates and tail docking of piglets at birth. The pursuit of higher stocking rates and greater mechanisation have contributed to widespread use of floor and hopper feeding systems. These methods make different operational demands on the stockman. Floor feeding, for example, may help to create cleaner lying areas than when hoppers are used but feed intake and consequently growth performance may be lower. Therefore, it is unhelpful to denigrate one method as opposed to the other and to condemn either as a 'fault' of intensivism. The enlightened operator seeks to minimise any disadvantages in the management of his particular system.

The economic demands of the modern pig industry have also led to other fundamental changes in production methods, and housing systems have been developed to accommodate such change. The prime example of this is the change to earlier weaning of pigs. Until the 1960s pigs were typically weaned at eight weeks. This then dropped to five weeks and now pigs are commonly weaned at one month or less in most parts of the world. Specialist accommodation to support the environmental and social needs of the pig weaned at around 5 kg has become widely adopted even if these needs are not universally accepted in detail.

Another key factor in the development of certain types of housing systems and production methods has been the degree of encouragement given to increased production by national governments. For example, grant aids available for building construction during the 1950s and 1960s in the United Kingdom resulted in a number of specialist enterprises developing on small acreages. This was a seductive route for those seeking to establish their own businesses and unable to secure large plots of land to farm ruminant livestock or for cropping.

The demand for cereal-based feeds generally means that the intensive pig enterprise becomes established within a reasonable distance of the source of such feeds in order to reduce transportation costs. In those countries which produce their own grains the pig business tends to be located close to the arable areas. Additional advantages of this accrue in the disposal of wastes and, where appropriate, the securing of straw bedding. Even where a pig enterprise is established as a part of a larger business, agricultural or otherwise, there is an apparent need to incorporate an enterprise large enough to justify the cost of the retention of a skilled specialist manager. This again tends to promote the selection of a reasonably large enterprise and in its turn means that low labour demands have a large impact upon the choice of housing system.

A typical pig enterprise meriting the description 'intensive' would contain certain specialist buildings. Many of these would incorporate part if not all of the floor in the form of slats and the discharge would be handled as a liquid or semi-solid. The need to reduce space allowances to limit costs means that individual dry sow penning is more likely to be seen. Loose or mixed penning requires larger floor areas to enable social order to be established without the more subservient sow being damaged. In more recent years the value of a specialist mating area has become acknowledged. This increases reproductive efficiency. The same necessity to increase output would mean that sows are likely to farrow within a specialist farrowing pen using some form of farrowing crate and a heated area for the pigs.

Weaned piglets are likely to be penned in housing which enables good control to be maintained over the temperature, at least in the lying zone of the pen. In the first two or three months of life the majority of piglets will be fed generous quantities of feed from a trough or hopper.

In later grower or finisher stages a wide range of feeding systems would be used from trough to hopper or floor feeding. All of these have certain merits or disadvantages.

This generalised description of an intensive unit conceals the obvious fact that a plethora of designs can be used to fit the outline given, underlining the important fact that there is no one 'right' way to accommodate a pig at each stage of life. This can be witnessed by the fact that some units fail to achieve good results with apparently good housing, while others at the lower end of the pig housing spectrum succeed in very basic buildings. Apart from the level of operational skill employed there is also the need to evaluate the criteria adopted in the use of the buildings. It must be stressed that a key lesson, slow to be accepted by would-be purchasers of a particular pig building, is that it is *not* good enough to adopt a given layout and set of dimensions and then ignore advice and recommendations on stocking rates, weight ranges of pigs and feeding scales used. In fact it is the operation as opposed to the design which may have the greatest impact upon results of both intensive and more traditional designs.

It is therefore impossible to define buildings as 'right' or 'wrong'. Each decision in the selection of a particular layout should come as a logical reaction to a series of well-thought-out decisions including legal, fiscal, welfare, environmental and operational considerations. Financial pressures are likely to mean that some, if not all, production phases incorporate some slatted and mechanical ventilation and feeding features. In addition some investors and some kinds of climatic circumstances will also dictate that there will always be those who ignore the modern approach to pig housing. Any book seeking a degree of comprehensiveness will need to define standards by which structures and layouts for the extensive methods of pig housing may be defined.

The route that this book will attempt to follow is to take an informed view of housing systems and to lead the farmer and manager through the essential decision-making processes required to arrive at the correct investment decisions and layout design. Such decisions must include some consideration of the criticisms, justified or otherwise, of all housing methods of pigs.

PIG HUSBANDRY AND WELFARE

The period from the mid 1970s to the present day has witnessed a growth in the number of groups of people seeking to influence the impact of agriculture upon the environment. Pressure groups are able to create newsworthy dialogue which criticises modern farming methods and, in particular, the more intensive systems of animal housing.

There has been considerable, largely academic, debate upon the topic of animal welfare in a number of countries, primarily in the northern hemisphere. The subject is one which engenders considerable emotion against the 'unnaturalness' of the systems employed, particularly the reduction of the use of bedding and the restraint placed, in some circumstances, upon pig movement. Now, somewhat belatedly, research is being encouraged into the design and operation of housing methods. This may yield a more acceptable return on capital employed and labour required with less dependence upon enclosed, unbedded methods. As already mentioned the development of modern housing methods has been driven by economic necessity and is the result in part of imprudent initial investment. Much of the disaffection of the farming community with the welfare lobbyists stems from the perceived failure of the latter to accept that livestock farming is a business and justifies a sympathetic consideration in that respect just as much as the needs of the animal itself demands respectful consideration by the farmer. To those with capital invested it is difficult to accept that there is anything intrinsically wrong with the profit motive, even if achieved by livestock production. Many pig producers and their staff feel a complete sense of mystification at the criticisms aimed at their 'modern' approach, especially when they spend many unsociable hours attempting to provide comfortable, congenial conditions for the stock in their charge.

However, much of the development of the 'intensive' house originates from the scramble for greater output from less staff rather than consideration of the animal's needs. The farming lobby must accept, therefore, that using these housing systems does make them obvious targets for criticism by those who fear lasting damage to livestock standards which may follow the acceptance of sow tethers, tiered cages, hen batteries and veal crates.

Some governments have already imposed legislative limitations upon certain housing systems. Others, as in the United Kingdom, seek to provide a code of recommendations for the design and operation of livestock buildings. However, it is the skill of the operator rather than the building design which is the true arbiter of animal comfort and tolerance. Many producers will see the attempts to ban particular designs or certain aspects of design as a failure to acknowledge the influence of the operator. In fact an operator's licence rather than the ban or acceptance of certain housing designs would be a more acceptable and reasonable approach by agricultural legislators. Some governments concentrate their efforts into providing standards for housing, stocking rates and feed space availability. Others, such as the United Kingdom, seek to influence the concepts underlying design and production systems, for example, the weaning age. However, the British Code of Recommendations (Ministry of Agriculture, Fisheries and Food Leaflet 702, 1983), does acknowledge:

Stockmanship is a key factor because, no matter how otherwise acceptable a system may be in principle, without competent, diligent stockmanship, the welfare of the animals cannot be adequately catered for.

No agriculturalist would seek to disagree with this premise nor other comments in the Preface of the *Code of Recommendations* that:

Pig husbandry systems in current use do not equally meet the physiological and behavioural needs of the animals. Nevertheless, within the framework of the statutory powers under which the Code has been prepared an attempt has been made, on the basis of the latest scientific knowledge and the soundest current practices, to identify those features where the pig's welfare could be at risk unless precautions are taken. The Code sets out what these precautions should be, bearing in mind the importance to the pigs of their total environment and the fact that there is often more than one way in which their welfare can be safeguarded.

These welfare code recommendations are issued under a 1968 Parliamentary Act (Agriculture [Miscellaneous Provisions] Act, 1968, Section 3(i)). Like similar legislation in other countries, they seek to ensure that no animal is subject to unnecessary pain or distress. In addition, in any case of dispute the recommendations would be interpreted as the standards by which an assessment of mismanagement would be obtained. In fact the concepts and recommendations in these codes or legislative measures cannot be criticised by the producer. The specific advice given on operation, safety provisions, temperature and ventilation guidelines and, particularly, stocking rates are all excellent yardsticks. It is not easy to conceive circumstances under which the advice given in these respects should not be followed.

The emphasis placed upon the increased technical understanding of more complex environment control equipment and automated feeding equipment is totally justified and should be paramount in all design considerations. In addition, as far as the United Kingdom is concerned, housing systems giving priority consideration to the use of straw bedding are strongly encouraged as, in the words of the 1983 code:

Bedding, and particularly straw, contributes towards the needs of the pig for thermal and physical comfort and satisfies some of its behavioural requirements.

The use of bedding immediately creates difficulty for the adoption of slatted housing systems because it is probable that blockages of the slurry channels and disposal equipment would result. At present there is little indication that housing manufacturers and their customers are enthusiastic about this advice, particularly for growing and finishing pigs. There are those who agree that a house should be designed sufficiently well for thermal and comfort needs to be satisfied without the use of bedding and that the behavioural needs of pigs at these stages to root and eat fibrous material is not at all well developed. Even so the pressure being placed upon farmers and building designers to plan for the use of bedding, particularly against the background clamour for the control of straw burning, is likely to yield a gradual change in emphasis for pig housing design in the future. This aspect will be dealt with in detail in a later chapter.

Farmers in general would be wise to accept that although recommendations given may be couched in moderate and sensible tones there is a militant body of vegetarian protestors in a number of countries seeking to limit farmers' freedom to operate their businesses in the most profitable way. Against this background it would be foolish to adopt systems contrary to public opinion even if this is only a minority view, without due regard to the alternatives. The selection of an unbedded or equally unpopular housing system should be the result of a great deal of careful thought and its merits should be able to be fully justified. It is also advisable that the pig producer pays attention to his ultimate customer's demand for a product produced in a wholesome, acceptable manner. Public opinion on what constitutes 'acceptable' may be subject to greater influence from the lobbyists than from farmer representatives. This illustrates again the need to acknowledge the opinions of those outside the agricultural industry.

Acknowledgement of the growing influence of the anti-farming lobby is important to the farming community in a number of countries. Failure to consider their opinions will only lead to some future confrontations and there is much that pig producers can do to deflect attention from themselves. Careful and diligent stockmanship is taken as read as there can be very few people who knowingly cause unnecessary suffering to animals in their charge. However, an area where pig farmers could help their case is in the positioning of buildings in relation to private dwellings and natural features. They should not fall back on obtaining appropriate planning approval as justification for some anti-social development. Furthermore livestock producers have not always taken enough care in planning their manure disposal to ensure that water-courses are not polluted and airborne stench is not created. If this aspect was tackled sympathetically, it would have a beneficial effect upon their public image.

The short-term effects of poor day-to-day management control have in the past incurred some justified criticism. Examples of poor control of service management leading to periods of heavy farrowings and over-stocking of buildings were all too common. No producer or his staff can take pride in being the cause of such circumstances. Ready-made monitoring and recording systems are now well developed, and these provide the pig unit manager with comprehensive information enabling him to avoid problems which have health and profit undertones as well as those of welfare.

Furthermore, paying attention to details such as draught exclusion around doors, windows, slurry chamber covers, drains, vents etc., is capable of yielding a large influence on animal comfort, behaviour and efficiency of performance. There can be no excuse for a failure to tackle these and similar operational problems.

In short, and as acknowledged in the wording of the British Ministry of Agriculture, Code of Recommendations for Welfare of Pigs, there is no difference between good, commercial considerations of stockmanship and those practices required to satisfy welfare requirements. If in addition rather more thought is given at the planning stage to waste disposal and public relations considerations the pig farmer need have little to fear from the activists.

FINANCIAL CONSIDERATIONS

Investment Appraisal

Investment in facilities for pig production should be subject to the same criteria as those used for any other business. That is to say the output achieved as a result of capital expenditure should be justified within an acceptable duration. The definition of 'acceptable' is however, not straight-forward because, like any other investment, it is subject to the personal circumstances of the investor who may take an individual view of his outlay.

Every investment should be capable of yielding a return on itself at least as good as that which might accrue from some safe source, such as investment bonds, gilt-edged securities and the like. In practice, investment in pig buildings and equipment has to be considered differently because of the lagged effect in physiological terms. A new farrowing house will have ten to twelve groups of sows through it in one year and is unlikely to yield the returns suggested in the same period of time as a similar investment in normal savings facilities. A greater degree of patience is therefore required in order to judge the merits of expenditure and it is for this reason that *cash-flow* considerations have become increasingly widely used for testing the merits of a proposed capital outlay. This approach is a valuable one because it also incorporates the build-up effect of stock, feed, labour and other costs before sales commence. Cash flow can also take into account the interest charges and the benefits of the capital outlaid, and any inhibitions on the business can more clearly be assessed.

By the very nature of the agricultural business—uncertainty due to climatic and seasonal effects and the incorporation of other enterprises on the majority of farms which keep pigs—it becomes even more complicated to reason the case for and against investment. Government decisions may also influence such judgements through taxation incentives or the very opposite. The vagaries of fiscal policies make predictions of income just as hazardous as do climatic influences. In addition, agricultural products, particularly pig-meat, are subject to cyclical movements in supply and demand and these are fuelled by political intervention for competing feed products. On top of all these factors comes the effect of interest rates which are of course subject to fluctuation and make predictions still more difficult.

All these problems tend to be more pronounced in pig production than many other sectors of business. Although there cannot be an immediate sale of product from an investment in livestock (unlike, say, light or manu-facturing industry) the relative speed of reproduction and growth means that in about one year supplies of pigs can grow sufficiently to cause demand to be exceeded with a commensurate drop in prices for pigmeat, and vice versa. Predictions of the timing of such cycles are notoriously imprecise.

Giving consideration to these confusing, arbitrary and unpredictable influences, how can a farmer judge whether an investment in buildings and equipment is prudent? Leaving aside whether a pig enterprise already exists or not, it is always sound thinking to consider the following likely factors in a conservative manner.

Productive efficiency should not be over-estimated. Reasonable levels of sow output and pig growth should be used taking into account the influence of reduced productivity from gilts and the possible impact of some disease problems.

Variable costs include stock, feed, fuels and transport costs. All these features have shown steady and sometimes rapid increase over the past twenty years. Their continued inflation is likely and should not be underestimated in any attempt to test the benefit of investment.

Finance charges have been variable over a short span of time in recent years largely as a result of international financial influences. A farmer may consider his finance charges in a variety of ways. If he has money available as a result of successful trading in some other aspect of his business, agricultural or otherwise, then he may believe it prudent to increase his investment in fixed equipment. However, it is rarely wise to consider an investment in intensive livestock buildings as an increase in the equitable value of a farm due to depreciation costs and varying opinions on the value of the buildings by any potential buyers of that business as opposed to the land itself. It is generally more prudent, where money has to be borrowed to fund capital investment, to budget for an amortised repayment to the loan agency as this takes into account the return of the capital sum borrowed and the loan charges which decrease with the staged repayment of the principal sum. The attitude towards repayment time will vary with the scale of the borrowings and the precise nature of the investment. Depreciation attitudes are also influenced by any taxation effect which may apply but it is reasonable to suggest that a flexible approach to the period over which buildings and equipment are depreciated should be taken into account for both changes in fiscal policy and the nature of the investment. Some types of pig building will more obviously and quickly present a measurable level of return including finance charges than others. Many countries have varying attitudes towards leasing of buildings and equipment and these may variously make consideration of this route more or less attractive. However, all these considerations and attitudes towards finance charges should be subservient to the major considerations that investment should *not* be made simply to follow a trend, fashion or whim. It should be made only on the basis of the potential increased viability of a business. Again, investment cannot be made on the basis of realisable value of land which, after all, if 'realised' would make farming impossible!

Non-financial considerations may also affect attitudes towards investment. Some arable producers on certain types of soil claim considerable benefits from farm manure returned to the land from a livestock enterprise even though such practices may be difficult to evaluate. A wider view of the sociological effects upon a community of the increasing mechanisation of agriculture may be taken. Pig production can help by offsetting the decline in the numbers of those both working and living in rural areas. A third consideration is that of a typical farming family where the next generation may have specific interests in pig production and where there is a need to increase the earning capacity of the business to satisfy the financial aspirations of more people. None of these factors are good enough to override the other main considerations for investment but they may influence the attitude taken towards the repayment time of loans and depreciation costs, making apparently less favourable circumstances acceptable.

Taxation is important because fiscal policies and concessions relating to investment change from time to time. The impact of fiscal policies must be considered in the light of that possibility. In other words investment, simply because it may seem attractive or expedient at the time, should not be allowed to overrule long-term assessment of the overall viability of a project. A good example of this occurred in the United Kingdom when a change in the tax laws was made in 1984 concerning write-off periods, making leasing of equipment and short-term tax relief far less attractive. It must be remembered that agriculture is the only major industry where most capital projects are still mainly funded by, and subject to, taxation on private or personal finances. The implication of imprudent investment or changes in government fiscal policy are therefore likely to be more acutely and directly felt. So is this introduction to considerations of investment over-cautious and gloomy? Choose any modest trading year in pigs—and these occur in every free country—and then the answer to that question is 'No, they are realistic'. Some would argue that 'pigs never look good—only poor and then worse' but such adages can be balanced by such equally sage observations as 'investment is an opportunity not a millstone' and so on.

So how can a judgement be made on whether to instigate a new pig business or extend the existing enterprise? There exists a series of yardsticks against which investment or the need to borrow monies may be assessed:

Is there a total commitment to a pig unit or more pigs?
If not, 'forget it' would be sound advice, as all livestock enterprises demand a willingness to become involved with the 'dirty end' of pig business.

How much confidence is there in being successful?
A proven track record of successful pig production allows more optimistic estimates of physical output to be made and makes the would-be borrower more attractive to the financial source.

How little can be borrowed to establish or extend a sound pig enterprise?
As will be stressed later, it is difficult to measure the difference in output between various building systems. To borrow or invest more money to have the 'latest thing' rather than apply greater levels of skill and borrow less, may simply make the achievement of equitable output difficult, and more hazardous if an unexpected trough in profitability arises. Remember that an increase in fittings etc., of 10 per cent based on preferred equipment or layout, must be capable of being measured against increased output or reduced running costs or better output per man (saving in labour). This is not to argue the case for selecting a sub-standard design or system but for a prudent consideration of the facts surrounding the degree of investment. Conversely, employing an additional person to offset inadequacies of investment may cost more than the saved amortised repayments when the total employment costs, which may include employers' costs, holiday and sickness periods and so on are taken into account. Again, no hard-and-fast rules can be laid down but prudent and careful consideration should be given. (See chapters on design and operation considerations.)

What about the cost of services?
It is easy to underestimate the proportionate costs of the non-productive part of a pig unit because it is so unglamorous and less interesting to contemplate. Differences obviously occur between any given set of circumstances but for all visible or waste facilities costed, it is wise to add a further 30 per cent for extension of services to and from the unit or building being planned. Requirements such as roadways, fences, soakaways, rainwater disposal and the like should not be overlooked.

What is the correct ratio of borrowings to assets?
To some extent the answer to this is influenced by earlier considerations mentioned in this chapter. However, as a broad measurement, few farm businesses can be expected to prosper where more than 30 per cent of gross

farm/unit output is being absorbed in financial charges regardless of the assets : borrowings ratio.

What else will suffer if investment is made or borrowings increased?
It is wise to consider whether other parts of the business will be hampered by a new or extended pig business. This could have repercussions on overall earnings and therefore place severe restraint upon personal drawings which may not be acceptable.

Who else is owed money?
It is always sensible to utilise all forms of credit including that available from trade sources. It is, however, a dangerous philosophy to use as 'security' for a loan for a new building, livestock which has already been used as 'security' for feed supplies.

Sources of Finance

Having carefully considered the pros and cons of an investment or a borrowing, from where might the money required be obtained? A list of sources of finance can never be absolute. The main creditors for fixed equipment and building loans in the UK are shown below. Similar sources may be available in other countries:

Banks are traditionally the largest source of capital financing. Bank managers and other potential lenders will normally expect to be presented with a plan which includes considerations and justifications of all the foregoing points in this chapter. Loans are normally agreed at a rate of some 2 per cent above current bank interest rate. Credit may be obtained in several forms from a bank. Briefly these are:

A loan. This is formed by using a fixed sum and a fixed duration of repayment. It is usually, though not necessarily, at a fixed rate of interest. These have an advantage in periods of volatile interest rates in allowing firmer budgeting of loan charges. The bank will not call in the monies lent on a straight loan if a difficult period of trading arises. A true bank loan, rather than a personal one, normally allows repayments to be negotiated at the outset. Loans for new buildings are sometimes called farm development loans and normally provide for up to ten years over which time their purpose should be substantially justified and are normally secured against some fixed or, possibly, floating assets of the farm.

An overdraft. This is the commonest form of loan. If used for buildings, it is wise to obtain a written undertaking from the bank of the agreed limit, interest rate and the terms under which the money is lent and repayments are to be made. Some banks will also charge a commitment fee so, even if the monies are not used or used to the full, their potential allocation to your business is safeguarded. Interest rates on overdrafts are calculated quarterly. The true interest rate is therefore likely to be a further 1 per cent above the 'base plus 2 per cent' calculation.

Because so much variation exists, wise farmers have learnt the value of presenting a good case well—and finding a well-informed, agriculturally minded bank manager to present that case to!

Finance Houses will advance money for hire-purchase type arrangements and will only normally seek security on large buildings or equipment sums. They offer easily obtained cash but interest rates will be high. So they may be attractive sources to the newcomer or those with limited financial resources available, but it is these very people who should be wary of over-extending their borrowings. Repayments are usually by monthly or quarterly instalments and monetary relief is normally available although hire-purchase repayments do have to appear on the balance sheet.

Venture Capital refers to funds which may occasionally be securable for short- or medium-term purposes, sometimes at favourable rates. Their main disadvantages are in locating the money in the first instance and limitations set as to their usage and the duration of the loan.

Leasing relates to certain items of equipment and forms of buildings. Container and some package-deal buildings are suitable subjects for leasing. It is attractive in that leasing fees do not appear on the balance sheet and so do not influence overdraft limits. Whatever capital allowances are available for taxation purposes can be taken into full account even from the start-up when no profits have been generated. Rental payments can normally be offset against tax. However, leased buildings do not normally qualify for grant aid and never become the farmer's own property. In addition the pattern of loan repayments is not normally negotiable and so commits the farmer to a fixed schedule.

Security is often difficult for a relative newcomer into agriculture to provide. Personal guarantors are widely-used sources of security in agriculture, normally family members or sleeping partners. In the UK the government funds the Agricultural Mortgage Corporation (AMC), which guarantees the applicant at his own bank to an agreed limit. This enables farmers to obtain loans when other security is lacking and the bank or finance house normally charge a lower rate of interest due to the decreased risk of their loan. However, the borrower has to offset this against the 1–3 per cent charge levied by the AMC.

Thus the rules for any investment are clear but their application is very much the concern of the individual and his financial advisor to justify outlay and borrowings. Whilst the hazards of predicting periods of relative profitability in pigs are apparent, it is wise to consider the cyclical nature of pig production and to avoid bringing a new business or extension into full production at those times when profits are at their lowest. To justify expenditure when pigs are in a 'trough' takes boldness, but to have that house and business in full production by the time improvements in margins occur is a wise aim.

The sequence in which the financial requirements are estimated does demand that a degree of planning and costing takes place prior to application for borrowing. However, study of this chapter before considering Chapter 2 may be sufficient to deter unsound investment.

Finally, if it is expansion that is being considered the decision should be subject to the same set of rules as a new expenditure. However in this case another question should also be asked! 'Could I expand output by becoming more efficient rather than increasing the scale of operation and investment in increased facilities?' Frequently the best answer is to say 'Yes' to such a question *and* to expand.

Chapter 2

PLANNING THE UNIT

SEEKING ADVISORS

GIVEN A SATISFACTORY appraisal of all financial considerations, the detailed planning of the proposed development can proceed. In practice it is usually necessary to establish in outline the building needs, so that the financial commitment can be assessed and the capital requirements and feasibility tested. However, detailed planning must be made before actual site clearance and instructions to proceed are placed. Badly conceived projects abound on many farms due to lack of foresight. Local Authority or Water Board approval may be withheld; future development such as amendments to feeding systems could be difficult, if not impossible, to achieve unless careful thought is initially given to the project.

Understandably many producers feel daunted by this planning process when faced with a list of considerations, of which the choice of building layout may be the easiest to decide upon. It is rarely suitable simply to copy someone else's ideas as every unit has its own specific needs. Even if the type of building is known in advance, its siting relative to other buildings, provision of services and future expansion must be individually considered. No two farm circumstances are ever completely alike. In planning, therefore, the first consideration should be the source of advice and assistance available to ensure that a comprehensive project plan is produced which will lead to a satisfactory conclusion. The degree to which assistance is sought will depend upon the farmer's own available time and expertise. If outside help is sought he should have already drawn up a list of his own requirements and preferences. These can then be debated and justified with the independent source of advice.

Pig buildings are primarily for production but their inter-relationship with other buildings or enterprises must be recognised. For example, if home-mixed feeds are to be used, the proximity to the grain storage/milling plant might be important. Positioning the buildings to permit convenient storage and disposal of effluent with minimum risk of nuisance could also be critical. These two factors illustrate the need to consider all aspects of a project at an early phase in the drafting of an outline proposal. The farmer should then ensure that any such development is compatible with other enterprises and local geographical and social conditions. For example, in a district where bedding is not available due to unsuitable cropping conditions it would be difficult to justify a pig layout based on straw bedding as this might be subject to expensive haulage charges.

Having considered the suitability of the enterprise and site to fit into the overall business (and this could include any existing pig facilities), it is necessary to assess the availability of labour required to operate the projected enterprise or development. This will include the routine work procedures. The farmer would also decide on the preferred methods of manure disposal, feeding and type of production involved. For instance, the production of weaners would require consideration of weaning age, the weight at which pigs should be sold, etc. These aspects would lead directly into considerations of the size of site required, quantities of feed and bedding needed, manure outputs and labour needs.

The would-be investor should consider these objectives in conjunction with the specific needs of his own farm and incorporating his particular personal preferences. His requirements should be based on the points listed below and presented to an advisor for his observations:

—The proposed scale of the project and type of output.
—The need to fit in with local circumstances and the constraints imposed by the site.
—Compatibility with other farm enterprises.
—Compatibility with pre-determined production circumstances such as feed, bedding and manure.
—Labour availability.
—Personal preferences.

Where might such advice be obtained? Many countries have state, or state-backed, advisory services. In the United Kingdom these are operated by the Agricultural Development Advisory Service, the Meat and Livestock Commission, Scottish colleges, etc. In addition, commercial organisations employ technicians, many of whom are experienced in buildings. There are also a number of qualified independent advisory sources on buildings, some trained in architecture or surveying and others with extensive production experience. Although the ultimate choice of professional assistance may well be dictated by the compatibility of the advisor's character with that of the client, the relative merits and disadvantages of the various sources of expertise may be generalised as shown below:

State-backed Advisory Services

- An excellent source of accurate advice on legal aspects and schemes such as planning approvals, grants, safety, water, electricity and gas board approval which they help to administer.
- Normally, a sound opinion is provided on site layout, services, drainage and security.

- It may not be possible to ensure that a full and sufficient discussion is obtained at every stage of the project's development.
- Working drawings may not be provided.
- There are not always staff available with experience of all design alternatives.
- Lack of commercial incentive may result in a slower response to the request for information and assistance.

Commercial Companies

- These can often offer the most up-to-date advice about new developments and on-farm experience.
- Sufficient time is rarely allocated to supervise all stages of project development.
- Some advice may not be entirely unbiased as they are often in the dual role of advisor and contractor/supplier.
- They are not always experienced in the planning and legal aspects of farm buildings.

Independent Advisors

- Adequate time is spent on the project to ensure that it will be effectively planned, approved, supervised and commissioned.
- Unbiased opinions will be provided and procedural work in obtaining approvals and permission for the work will be undertaken.
- The number of professional architects or surveyors who are experienced in many of the specialist design features in pig buildings is not great and, by the same token, the technically competent non-qualified advisor would rely upon others to interpret the proposals and to produce drawings.
- An independent advisor charges fees which vary from six to ten per cent of the cost of new works to thirteen per cent on existing buildings. Whilst such expenditure can be very well worthwhile because satisfactory work and smooth completion of all planning phases are ensured (and such costs may be eligible for a grant), it is vital that the client is satisfied that the chosen advisor is competent in all aspects.

Having studied the previous section, the pig producer planning a new building or unit should now conclude that:

- It is always worthwhile to consult the local government advisory centre to check on legislative factors.
- The choice of advice will depend upon his own capabilities and the type of work needed. It may well be acceptable to brief a qualified architect or surveyor to conduct application for grant, draw up plans, etc., based on sketches created by himself or another. An independent advisor could then be used to study and comment upon the plans to ensure that the critical points of design detail and equipment specified are correct and up to date.

- It is always worthwhile inspecting various layouts and systems, as well as specific items of equipment, before initiating plans.
- The final decision on expenditure, siting and building design will rest with the producer himself as there are relatively few items where alternative options are not acceptable. It is largely a matter of choosing the most suitable (or sometimes the least *un*suitable) course of action.

Do-it-yourself

Turning to the do-it-yourself approach, as an alternative to seeking outside opinion; what procedure might the farmer follow in making his decisions? It is important to divide these wherever possible into two major areas. Those pertaining to the site and strategic matters of services etc., and those which relate to specific design. Whilst these two are to some extent interdependent it is best to set aside specific aspects of design until thought has been given to the site as a whole. The farmer, and possibly his chosen advisor, might work through a check list which should help to eliminate a number of design queries. Such a sequence of planning is discussed below and it presumes that the financial considerations (see Chapter 1) are compatible with the projected development. Whilst the procedure and considerations set down are assessed from the point of view of a new development, little modification would be necessary to apply these same points for the addition of a new building to an established pig enterprise.

At its most basic the do-it-yourself approach might be perfectly acceptable given careful pre-planning using simple free-hand sketches. Indeed, a certain amount of such work is always worthwhile in order that some options can be eliminated and better use of professional persons' time made at some later point.

CONSIDERATION OF THE SITE

Planned Output and Scale

Summaries of the economics of pig production generally indicate that the most profitable units are those which are integrated breeding and feeding enterprises and which market pigs at relatively higher carcass weights. However, an integrated pig enterprise does call for a greater variety of more specialised buildings and operator inputs, and such diverse needs and skills may not be available to every producer. There is little doubt that more supervision will be required in a breeding enterprise where each day it is critical to check for oestrus and mating sows as well as improving piglet survival rates. Whilst close monitoring of specialist finishing enterprises will help to yield better results, they do allow for less regular supervision than breeding units. Heavier marketing weights create needs for more and larger buildings as well as extended cash-flow requirements, and these may also influence decisions on output.

The economics of scale cannot be defined in simple terms. For example, a pig unit of relatively modest numbers could be part of a number of enterprises operated and could also constitute part of a large business. In short, a large pig unit will only be successfully 'managed' where the proprietor or his appointee has time to exercise close control. It *is* relevant to point out that there are few advantages of scale from a productivity standpoint. Slight losses in absolute performance which may occur on large sites might be compensated for by reduced labour and machinery costs per unit of output. Thus, scale of organisation, of improved buying and marketing power may be attractive to the investor with access to resources and given suitable consideration of other points which follow. Scale will also be dictated by other factors such as the degree of mechanisation chosen. A *preferred* unit size rather than an absolute one has just as much to commend it.

Topography of the Site

The traditional siting of units was based upon protection from prevailing winds. However, the modern pig unit may have other primary considerations dictated by logistics. Nevertheless, the microclimate in which it is proposed to site the building should be carefully considered.

Naturally ventilated buildings require reasonable shelter by windbreaks but for warm weather venting they also need a reasonably open aspect. A sheltered hill-top may be preferred to a valley bottom because the higher site will be more likely to ensure natural ventilation without a tunnel effect. Sensible use of any slopes on the site to effect good drainage and to ease stock loading and inward goods unloading should also be considered.

Siting of individual buildings will depend upon climatic conditions as well as the style of building chosen: for example, dual or monopitch; fan or naturally ventilated. In the northern hemisphere a north:south axis may be preferred with a monopitch structure facing the warmest sector. In warmer climates siting to expose the building to the greatest influence of any prevailing wind regardless of the axis line may be the primary consideration.

Constraints Imposed by the Proposed Site

Although the slope on a site may be used as mentioned above, care must be taken to ensure that it does not create exorbitant site preparation costs. This must always be contemplated against a background of the possible future expansion of the pig enterprise. Buildings are almost always cheaper to build on a level site and foundation works comprise a higher proportion of the total cost of pig buildings than they do in many industrial structures. So careful planning can ensure that the site contributes to, rather than adds to, the cost of the proposed works.

If the site is next to urban development or public rights of way special consideration must be given to both planning (see Appendix 1) and security. It is better that a modern pig unit development should offer access to buildings from within the farm perimeter only. Security will also be helped if the unit is sited reasonably close to the farm dwelling, providing control over intruders as well as fire or other potential hazards.

Another planning factor to consider is the position of any home-mixed food source. Having this reasonably close to the building development would allow cost-effective use to be made of delivery augers and may cause some compromise to be made in the choice of layout. This also applies to slurry or solid muck disposal. The preference to store the latter in an unexposed position and to avoid the need to pass through major highways or areas of habitation might well rule out some sites which are otherwise suitable. Certainly, these points constitute an important part of the discussion phase of any pig development.

Whilst local authorities are the final arbiters of aesthetically acceptable siting, farmers would do well to ponder the facts connected with the appearance of farm buildings carefully. Not only might such considerations assist in creating a more harmonious relationship with local inhabitants but they will increase the chance of planning approval being granted. Both the shape of modern farm buildings and the materials used tend to be antagonistic to natural surroundings. Consideration given to screening the site with trees and hedging or retaining existing screening, using earthworks, siting electricity supply poles cleverly, using profiled cladding, choosing sensible colours and tinting otherwise light-coloured roofs can all contribute to the improved appearance of a building or buildings.

Provision of Services

In attempting to avoid disturbance of existing amenities or human habitation, some producers have planned to develop more isolated new sites only to come across different types of difficulty with services.

For example, a unit on a remote new site may incur a mains water provision charge which makes such a choice of position unacceptable. The average pig will require about 18 litres (4 gallons) of water per day for feeding, plus an allowance of 5 litres per day for cleaning purposes. The size of the water main and on-site storage capacity must be considered. It is always preferable to provide for on-site storage of 72 hours supply to safeguard against unforeseen circumstances. This means that a unit holding 1,000 pigs would require supply lines and a storage tank holding 69,000 litres or 69 m^3 (2,436 cu. ft or 15,000 gallons).

The supply of electricity may also pose a problem when choosing a site. The use of infra-red heaters for young litters and fan ventilation for growing stock, in addition to lighting, can easily create a need for some 200 kVA. This excludes surge loading such as that created by mill-and-mix units or continuous loadings from slurry aeration/treatment plants. Table 2.1 illustrates possible loadings created by such equipment. If ample load capacity is not readily available costs of providing or up-rating the electrical supply must be borne in mind. Sometimes it is feasible to separate heating cost loadings by using bulk propane fuel to overcome the problems of inadequate electricity supply.

Table 2.1   Electrical energy consumption

Mill-and-mix (cereals)	30 units/tonne
Wet feed mixing	3 units/tonne of dry feed
Wet feed conveying	8 units/tonne of dry feed

Rainwater drainage should always be considered at the planning phase. The aim will be to make on-site operations more convenient and also to keep such water away from foul water or slurry drains, thus reducing the volume of liquid to be disposed of.

Access to Buildings

The prime need of keeping working locations reasonably close to each other to reduce the time spent passing from one to the other and moving feed and stock must be balanced against the need to provide the appropriate vehicular access to the unit and specific buildings. The delivery of bedding, removal of manure and manoeuvring of slurry tankers all play their part in determining the position of buildings and the distance between them. It must be remembered that height and width for tractor-operated equipment and turning circles for tractors and their attachments should also be taken into account.

Distance between buildings should also be influenced by improvements to air flow particularly in naturally ventilated structures. In temperate climates the eaves-to-eaves distance should be no less than twice the eaves height of the taller building. In warmer regions a good working recommendation is a 10 m (30 ft) eaves-to-eaves gap to make possible an increased exhaust of air. Closer spacing of buildings may not interfere with the satisfactory working of fan-operated ventilation systems. However, care must be taken where mixed ages of stock are kept, and where exhaust gases may be drawn into the adjacent building with airborne organisms and temperature/humidity gains leading to potentially more difficult operating conditions.

Table 2.2   Basic dimensions for tractors and their implements

	Height clearance (m)	Length (m)	Turning circle (m)	Turning radius (m)
Small tractor with cab	2.4	3.0	1.8	6.4
Medium tractor with loader	3.3	5.8	7.0	10.0
Tractor and slurry tanker (7,000 gallons)	3.75	8.0	7.0	10.0

(After: Weller)

Disease Control

There is little doubt that disease security is now more important in unit planning due both to the influence of health standards upon the economics of production and to the increased hazards of larger units and increased geographic concentration of pigs. Positioning the unit at a distance from other enterprises is advisable. There is now doubt that anything less than 5 km (just over three miles) can be considered safe against air transmission of common respiratory diseases such as enzootic pneumonia. Other epidemic diseases, such as foot-and-mouth disease, are capable of even greater climatic spread. The relative risk has to be assessed and most producers may have to reconcile themselves to some degree of risk from airborne transmission. Serious thought should be given to the restriction of any contact between vehicles and equipment belonging to the unit and outside vehicles.

The provision of a 'clean' unit or area and the fencing out of feed, livestock vehicles and their personnel should be carefully considered (see later chapter). Much can be achieved by siting bins and stores so that they can be filled from outside the production security boundary and by positioning the loading zone carefully (see plate 1).

Naturally ventilated buildings require protection against the risk of birds acting as vectors of disease. Netting can be effectively positioned over vents and open sides of structures so as not to interfere with effective ventilation or

Plate 1 A unit secure from man, beast and, as far as can be arranged, disease. Note peripherally sited bulk feed bins and reception area (on right).

house operation. These must not inhibit the intended air flow in buildings and mesh sizes of 15 mm are a suitable compromise.

Future Development

There are many known instances where the most convenient development of a site has been precluded because future developments in production were not foreseen or planned for. There may exist unalterable limitations against future expansion, such as farm acreage. Where no obvious constraint is apparent the safest approach is to make provision for doubling the level of production now being planned. Consequently allowance should be made for the production of twice as much waste and transport of materials to the unit and the use of twice the water and fuel.

Size of site is not the only potential constraint. The position of buildings should also allow for some future development without excessive compromise on unit operation, pig movement, etc.

ASPECTS OF DESIGN AND LAYOUT

Considerations of the actual system of production chosen and specifics of house design are inter-woven with the choice of site and deployment of facilities on a site. Basically, the choice of production system will depend to a large degree upon the amount of mechanisation to be used for manure disposal and feeding, which in turn will affect labour needs. The greatest single influence upon the dimensions of a building and its juxtaposition with others is undoubtedly the choice of manure disposal. It is therefore logical to begin considerations with the selection of a manure disposal system. However, given no constraints caused by the site itself, it is perhaps a good idea to review the more important factors in the layout of a unit.

Choosing the Unit Layout

Occasions which provide a free choice in the siting of buildings are not common, because of many of the points discussed. Just how can the producer set about choosing the best option even if a degree of compromise has to be applied?

Health security is very much improved where vehicles do not have to proceed past production facilities. This applies particularly to livestock vehicles which pose the greatest potential contamination risks due to their regular contact with other sources of stock. Pig collection and delivery points should be nearest to the main road access and obviously this should dictate that finishing houses should also be situated here. The opposite is typically the case. On an integrated breeding/feeding unit, the farrowing quarters are often placed nearest the entrance, presumably to be convenient for evening inspection. This is wrong for logistical reasons and for health security.

There is sense in laying out the various buildings so that there is a logic of

Logic chart in pre-planning phase

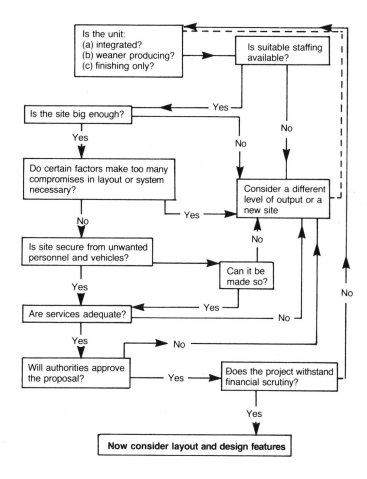

pig flow around the site. For example, if finishing houses are positioned by the entrance/exit, grower houses would be nearest to them with weaner, farrowing and dry sow houses still further away in that progression. This can considerably assist operator convenience because those who have responsibility for farrowing areas do not have to walk through other production facilities to conduct their routine. The stores for the delivery of bagged feed and for light goods and equipment should be sited adjacent to the entrance and on the perimeter of the unit. In addition, high priority should be given to the positioning of bulk feed bins inside a secure fence for filling from outside. All the livestock buildings should be enclosed within a fenced area to give total and complete control over who and what enters the production area. It is also advisable to arrange for manure and slurry collection to be carried out from outside the perimeter of the production area if possible.

Service buildings, such as office, mess room and isolation/acclimatisation premises should be peripherally positioned to afford a convenient barrier between the outside and the production area. It is particularly useful if the staff area has a clear view of the entrance and a bell or siren is fitted to ensure that visitors can draw the attention of personnel on arrival at the unit. Considerable thought is often given to wheel dips and the positioning of the inevitable *keep out* sign. Security against crime and disease are both improved by having a well-planned layout surrounded by a good fence and operated to a strict routine which excludes lorry drivers and all other non-essential personnel. Fresh manure is an active vector of disease and any risk of its transmission into the unit should be minimised.

This again underlines the need for all planning to be conducted against a backcloth of possible expansion and future development. Indeed, if an extension is being planned and some facilities already exist it is best to consider changing the use of an existing house or houses rather than create a difficult layout for operators by adding an additional building. These comments are highly relevant for larger units where convenience of operation can contribute to improved efficiency.

Choice of Manure Disposal System

After two decades of increasing reliance on slatted floors and the disposal of manure in the form of slurry, designers and farmers in the United Kingdom have been forced into reviewing this in the light of the Welfare Code recommendations. There are also increasing pressures to utilise straw rather than to burn it in the fields.

The advantages of a bedded system do not rest with the difficult-to-measure view that pigs 'appear more comfortable annd content' when allowed access to straw bedding. Straw can allow less sophisticated systems to be used, acting as a buffer against a lower air temperature than that recommended. In addition, the regular removal of strawy manure generally creates far fewer odour problems and it is cheaper and more convenient to store the manure in a heap for eventual spreading. Straw also has an absorbent capacity which provides a greater safety element against any potential pollution in the event of a spillage.

The trend away from the use of bedding implies that there are disadvantages in its use. The cost per square metre of building may be lower and there may be less equipment needed to store and treat solid manure. However to offset against this is the need for larger buildings to allow for some mechanical removal of the manure. Openings are required to get straw into the buildings and areas of concrete are needed outside the buildings to permit easy removal. Storage space is required for its protection and there will also be an increased daily labour requirement, not to mention the acute peak of labour required after harvest to cart the straw home.

For calculating output of solid, straw-based manure the annual straw consumption of a farm would have to be calculated from the table below.

Table 2.3 Estimated quantities of straw needed for a bedded pig system

Class of pig	System	Straw/pig place/year
Dry sow	Bedded lying area and scraped yard	1 tonne
Sow and litter	Cleaned daily	750 kg
3–10 weeks	Kennel and scraped yard	80 kg
Finisher (25–90 kg)	Kennel and scraped yard	200 kg
Finisher (25–90 kg)	Deep straw	350 kg

Fortunately areas where land is not given over to cereals are often used for grass where it is possible to arrange continuous slurry disposal, whereas land in an arable area is only available seasonally.

Although it is a fair claim that slurry-based systems have been evolved without giving encouragement to innovative development of improved methods of straw and solid manure handling, the thought of having to 'move in' straw and 'move out' solid manure for every pig on a typical, large-scale modern enterprise is an alarming exercise in logistics. The disposal of slurry from a large unit can create similar problems. The choice is therefore more likely to be determined by availability of straw, the desire to use solid manure in a cropping programme, proximity to urban development (solid muck being more tolerable in such circumstances) and the farmer's own preference. Proof that pigs perform better when bedded is difficult to obtain and the decision will be based more on the individual circumstances and legislative prompting than by any finite evidence of pig efficiency.

Slurry storage is capital intensive and the expenditure rankles with farmers because it is an unproductive outlay. The idea of using the store as a latent source of energy to produce recoverable heat or to generate methane for providing power to drive an electricity-producing generator is most seductive. However, these ideas tie up even more capital and few such plants have yet to prove cost-effective in the northern hemisphere.

It is probable that pressures will increase for livestock producers to site units away from urban areas and for slurry and manures to be moved in a fresh stage (before becoming anaerobic and smelling offensively) to a store or to be spread away from the farm. Even normal pig smells are likely to prove less acceptable in the future. In parts of Western Europe it is already obligatory to 'clean' exhaust air from fan-ventilated piggeries sited within 1 km of a village or town.

Table 2.4   Estimated outputs of excreta produced by pigs

Class of pig	Body weight range (kg)	Ave body weight (kg)	Estimated excreta produced (litres/day)	Moisture content (%)
1 dry sow/adult	–	140	6.0	90
1 sow & litter (to 3 weeks)	–	170	15.0	90
1 piglet 3–8 weeks	5–20	12	2.0	90
1 pig place (dry feed)	20–90	50	4.0	90
1 pig place (wet feed) (water to meal ratio 2.5:1)	20–90	50	4.0	90
1 pig place (skim whey and swill)	20–90	50	14.0	98

Most countries prohibit the discharge of raw or treated farm wastes into water courses. In addition careless application of manures which create a pollution hazard or a nuisance to local inhabitants are also subject to legislation in the United Kingdom and elsewhere. The potential increase in nitrate levels in naturally occurring water resources from excessive manure application to fields may also be a problem if guidelines on application rates are not heeded. Certain methods of spreading, particularly those where small droplets of liquid are sprayed into the air can increase the risk of disease and aggravate odour problems, although storage for two months or more prior to spreading would reduce such risks.

Soil and climatic conditions may have a considerable effect upon the way in which manures—particularly slurry—enter the soil after spreading. Run-off, percolation via cracks into land drains and thence into water courses can occur in clay soils which may be too wet or too dry at the time of application. In addition if the soil is too wet and the soil temperature below 4°C (40°F) normal breakdown by micro-organisms may not occur.

There is no doubt that valuable manurial values exist in pig wastes which also have a proven record as a good soil conditioner. The relative inconvenience and cost of spreading frequently means that neighbouring farmers are loth to value piggery wastes as a substitute for inorganic fertilisers particularly as, in the case of slurry, precise values are difficult to determine. This is generally due to the variable system of producing these wastes and their dilution caused by added water and spillages. To avoid problems with nitrate build-up, run-off problems and the contamination of pasture with copper which is commonly fed as a growth promoter to finishing pigs, no more than 50 m³/ha (4,500 gallons/acre), or 5 mm per application within a three-week interval should be made.

Although it may be possible to estimate the reductions in purchased fertilisers which can be made when slurry or solid manure is applied, the most important factor is satisfactory disposal of the slurry rather than the particular value of organic manuring. The slower breakdown and release of plant foods from solid manures means that the rate of application is likely to be governed by the loss of soil incorporation or subsequent pasture growth.

There are a number of potential problems in slurry handling. The following points may be helpful in reviewing these and may also help to determine the choice of manure disposal system:

Legislation. There will be increasing pressures to use bedding, reduce odour levels and minimise the contamination of natural water courses and underground reserves.

Sensitivity of the site. The direction of prevailing winds should be closely observed. The position of the proposed site may be too close to human habitation.

Design of manure disposal system. It is possible to reduce odour levels by good handling systems. These should allow for regular removal from the vicinity or on-site treatment.

Management. The maintenance of a clean, tidy unit should be possible and manures should not be spread at times and in conditions when odour problems are increased. Ways and means of avoiding nuisance should be studied.

Feeding Systems

The second major influence upon choice of system and building layout is the method of feeding. Whilst this is particularly true of growing and finishing pigs, automatic feeding of other classes of pigs can also have a major impact upon design and, therefore, dimensions of a building. Various feeding systems are discussed in some detail in Chapter 5. The following three factors have considerable effect on choice of design and layout:

- Hand feeding. This requires the provision of wider passages.
- Floor feeding. This allows a greater stocking rate and less restriction on pen shapes.
- Trough feeding. This imposes constraint on pen shape to permit adequate feeding access.

Calculating Space Requirements

From a cost viewpoint space everywhere in the unit is at a premium. However, the provision of insufficient space not only makes operations more difficult but can cause behavioural and performance shortcomings in the pigs themselves.

It is necessary to make certain assumptions about pig performance and operational management in calculating the number of pens required. Theoretical calculations must always be assessed against estimates which are realistic, not optimistic, because an unexpected delay in the despatch of some pigs due to climatic conditions or a breakdown can impose acute space difficulties. (Detailed standards of pen size are given in Chapters 7–12.)

The base point chosen may influence the calculation. For example the following criteria might be used:

—A preference to keep a given number of sows.
—The targeted number of farrowings per week.
—A planned output of weaners or finished pigs per week.
—An intention to produce a particular weight of pigmeat per week.

All these base points could give the same answer but this is only possible if the assumed performance standards are manipulated to do so.

The 'set number of pigs' concept is useful in defining the scale of the operation. However it is not really a good starting point because such a base may well impose inconvenient constraints upon the size of the farrowing rooms required. It is far better to start with an idea of the approximate weekly flow of pigs or weight of meat to be produced and to work back from that. Another point worth making here concerns two items which are frequently omitted from pen requirement calculations: firstly the time taken to 'turn round' pens, i.e. to completely empty pens or house sections and carry out between-batch hygiene routines; secondly the need to provide suitable acclimatisation quarters to accommodate replacement breeding stock. Too often, incoming stock are penned in facilities which are unsuitable, compounding the problems created by transport and a change of farm, and leading to a disappointing subsequent breeding career.

Table 2.5 includes the various factors which need to be considered in calculating the pen numbers required for an integrated breeding and feeding herd. A weaner production unit or a specialist feeding site may be assessed by using the relevant sections only. Any estimate is limited by biological aspects such as period of gestation and acceptable rates of growth. Also included is the sequence of planning necessary together with some suggested standards for various production systems.

An all-in/all-out policy usually requires a greater number of pens than a continuous flow system because discrepancies in performance tend to create slight overlaps. However it is difficult to maintain hygiene standards during the operation of continuous flow systems, and the air space will be shared by pigs of a widely disparate age/weight range. The consequences of this may not be serious with adult breeding stock, if sensible acclimatisation procedures are used for incoming gilts or finishers in later stages of production. There is good sense however in sub-dividing the farrowing, weaner and grower stages into weekly or, possibly, bi-weekly batch sizes. It is also recommended that the farmer should avoid starting with a rigid, absolute idea of herd output or

Table 2.5   Factors influencing calculation of pen numbers

Information required	Suggested standards of performance
Flow of pigs per week to be sold	As targeted
Number of pigs weaned per litter	Mature herd—9.5
Litters per sow per year	3 week = 2.35 4 week = 2.25 5 week = 2.15
Weaning age	As scheduled
Weaning-to-service interval	Plan for 10 days
Boar:sow ratio	Plan for 1:20
Estimated sow replacement rate	Plan for 40 per cent per annum
Selection or purchase to service interval for replacement gilts	Allow 7 days/gilt for every 5 kg below 110 kg at selection/purchase
Period of occupation of weaner penning	As planned
Estimated growth rate in weaner penning	Weaning to 8 weeks—350 g/day
Period of occupation of grower house	As scheduled
Estimated growth rate in grower house	8 to 13 weeks—500 g/day
Weight of pigs sold: occupancy of finisher penning	As planned
Estimated growth rate in finisher penning	13 weeks to slaughter—650 g/day
Between-batch cleaning and resting at each phase Continuous flow or all-in/all-out policy	see text

size. The numbers of pigs kept should be adjusted to give a convenient number of farrowing and weaner places per room. For example, a 100-sow herd weaning at four weeks gives from table 2.5, 100 × 2.25 ÷ 52 = 4.3 litters per week on average. It would be better to keep 92 or 116 sows in a herd to be weaned at four weeks. This would give a mean of four or five farrowings per week allowing easier pen allocation.

The period allocated for 'between-batch' hygiene is largely dependent upon the design of the building. For example, an unbedded house using totally slatted floors could be expected to be dry enough for re-occupation 24 hours after being washed clean. A solid floor system really needs 72 hours to dry properly. In addition, the average period of occupation of a farrowing house by sows prior to farrowing should be taken into account and included in the between-batch calculation. As a general rule, although slightly more expensive to provide, it is easier to organise between-batch cleaning with smaller rooms.

It is possible to calculate the penning needs of a pig enterprise using table 2.5. An example of this is shown below. It allows for adaptation to varying circumstances.

Calculating the Herd Size and Penning Requirements of a Pig Unit

Example:	A farmer aims to market around 80 pigs per week at a mean of 100 kg from a breeder/feeder enterprise, weaning at around 3 weeks of age.
Step 1 Weekly Output	To produce approximate 80 pigs per week would require 80 ÷ 9.5 pigs per litter = 8.4 litters per week. Therefore, as recommended, the farmer will choose to either farrow 8 or 9 sows and gilts per week and will either modify his estimated weekly output to 8 litters × 9.5 or 9 litters × 9.5, i.e. 76–85 pigs per week, or will adjust his expected output per litter. For this exercise the output assumed is *8 litters per week*.
Step 2 Farrowings per week	To achieve an average of 8 farrowings per week with 3-week weaning it is estimated that 8 × 52 = 416 litters per year will be produced and, at an annual output of 2.35 litters per sow that 416 ÷ 2.35 = *177 sows* will be kept.
Step 3 Farrowing Pens	If sows suckle litters for an average of 21 days and it takes 2 days to clean and dry each farrowing room of 8 pens and each sow spends 5 days in the farrowing room prior to farrowing, each farrowing room has a 4-week cycle. (It may be desirable to allow a longer period for cleaning and drying.) Thus, 8 × 4 weeks = *32 farrowing pens* (preferably in 4 rooms), are needed.
Step 4 Dry Sow/Gilt Pens	The number of dry sow places needed is 177 less those in the farrowing house at the lowest time (i.e. during cleaning), *plus* maiden gilt places. Thus 177 − 24 = *153 dry sow places* are required and some of these may be provided in the form of a service area. In a group system sows would be penned in groups of 4–8 to suit the weekly output. In a 177 sow herd with a 40 per cent replacement rate, 177 × 0.4 = 70 replacement gilts will be needed in a year. Assuming the gilts are purchased at 90 kg (see Table 2.5) they will be on the farm for 110–90 kg ÷ 5 kg gain per week = 4 weeks prior to service at which point they will be considered as part of the dry sow herd. Thus, 70 gilts ÷ 52 × 4 = *5–6 maiden gilt places* are required. Thus, 153 + 6 = approximately *160 maiden gilts weaned, sows and in-pig sow places* would be scheduled.

When allocating pre-service/acclimatisation or isolation penning, it is recommended that space for the penning of cull sows/resident stock is made alongside the pens to be used for new stock to achieve adequate acclimatisation. In addition such penning might be arranged to include space for sick or injured pigs removed from the main herd.

Step 5
Boar Pens

Using a 1:20 boar:sow ratio, the boar places needed would be 177 ÷ 20 = *9 boar pens*

It is worth making an allowance for boar acclimatisation penning in the 'utility area' (see Step 4), as a boar strength of 9 demands a replacement of around 4 per year, and given a 28-day acclimatisation period, there would be a pen occupied by a young boar for a considerable portion of the year, and this warrants the provision of a suitable facility.

Step 6
Weaner
Places

Because weekly output is estimated at 76 pigs per week and each pig is to stay in the weaner accommodation for 4 weeks (including 2 days cleaning/resting period), there is a need for 76 × 4 = *304 weaner places*, preferably in 4 rooms.

Step 7
Grower
Places

If pigs are to be grown in just under 7 weeks to 36 kg before transfer to the finisher pens, using the growth rate standard shown in Table 2.5, it can be assumed that pigs might average 14 kg on removal from the weaner area (given an approximate 5 kg weaning weight).

Thus, there will be an occupancy need of 36 kg − 14 kg ÷ average growth rate/day (500 g/day—see Table 2.5) = 44.

Thus, 44 days (6.3 weeks) × 76 pigs per week gives the calculated grower places needed. If no between-batch hygiene is practised = *480 grower pig places* are needed.

Step 8
Finisher
Places

With an average growth rate of 650 g per day and an average liveweight advance from 36–100 kg in this section each pig will occupy a finisher place for 100 − 36 kg ÷ 650 g per day = 98 days or 14 weeks. Because there are 76 pigs per week there is a requirement, with no allowance for between-batch cleaning, of *1,064 pig places*.

It is worth noting that, given a static herd size and well-regulated farrowing and marketing schedules, it is difficult to arrange for completely efficient use of pen space, so some additional allowance of pen space may be made to allow for the unforeseen circumstances and/or erratic growth of pigs within the weekly batches. Of course, such conditions might be mitigated by mixing of 'stragglers' or the marketing of a percentage of pigs at a lighter weight but these solutions both have the tendency to reduce margins per pig and such a factor makes the provision of some additional, possibly smaller, penning to accommodate the residue of a week's output easier to justify.

Thus, this example illustrates how a modest set of performance standards can be used to plan a convenient herd size and the penning that such a throughput would demand. Adjustments to the performance figures used, particularly if higher performance norms are specified, will call for higher standards of management to be applied and will both make the consequences of any shortfall more serious and less acceptable due to the reduced margin for error. These calculations should be given most careful consideration, and the growth and output standards given much thought before the schedule is adopted as a model for any building development.

Safety

An important consideration is the security of the stock under circumstances of systems failure or accident. Modern pig units tend to be larger than those used by previous generations of producer and have a greater reliance upon automation. It is important that the potential hazards implicit in a modern pig enterprise are acknowledged at the planning phase. Automated heating and ventilation systems must have incorporated fail-safe mechanisms which protect the stock in the event of power or systems failure. Ideally, this would be coupled to an alarm system warning operators of a fault even during rest periods. (Reference to stand-by generation supplies is made in Chapter 3.)

Escape hatches for stock in the event of fire or severe structural damage should be provided in enclosed systems. Removable wall sections are usually incorporated to increase the possibility of escape. It is important to stress the regular checking of standby or safety equipment or systems. They are only as good as the state of readiness in which they are maintained.

PLANNING THE WORKS

Having considered aspects of the site and systems, and having taken the advice of trusted specialists, the farmer now has to consider which method of construction he will use. The three main choices are discussed below:

Do-it-yourself Construction

This approach calls for a great deal of supervisory input by the farmer or his appointee, whether traditional materials or prefabricated parts are used. Considerable cost savings may be the reward for this even though the actual works might be subcontracted in total or part. However, it is essential to have accurate drawings and specifications drawn up for local authority use. A large range of quotations should be obtained and compared, and great care needs to be taken to ensure that all components required for the construction arrive on site at the required time, thus avoiding delays.

The use of do-it-yourself construction need not imply any compromise in building quality or performance. Precise provision of the preferred facilities can be achieved if the farmer is prepared to allocate sufficient organisational and supervisory time, or alternatively pay someone else to do so.

Using a Contractor

Although using a contractor to construct the building to an agreed specification relieves the farmer of certain organisational demands, it still allows all preferences in design to be applied. The costs may be higher than the D-I-Y approach because the contractor will apply a margin for profit. Regular on-site supervision by the farmer is very important.

The Package Deal

In many circumstances the farmer may be satisfied that a manufacturer can supply a building or buildings which match his requirements, particularly if he has already visited erected houses of the type proposed. However, allowance must be made for the skill of the operator concerned when judging such buildings. The purchase of a package deal removes the burden of organisation, and this convenience as well as the speed of erection of the building are often an irresistible attraction to a busy farmer.

In practice, a fourth option is available whereby part of the building or unit is constructed by the farmer and part by an outside contractor or package-deal manufacturer.

Planning a new development

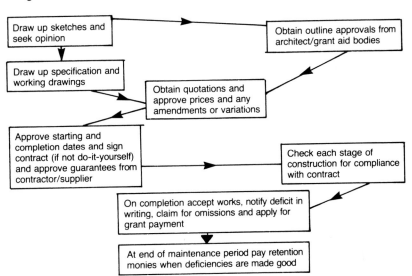

Conversions or *extensions* of existing farm buildings frequently bring their own problems. The current production programme has to continue unobstructed by the modifications and this requires careful planning particularly where a building is to undergo a change of use.

The chart on page 33 is an aid to the various stages of planning a new development.

Having chosen the type of building, the site layout and the method of construction it is necessary to account for the time taken to obtain planning approvals and to erect the buildings. Wherever contractors or outside personnel are involved it is essential that a completion date becomes part of the contractual agreement. It is necessary to plan stocking schedules to avoid expensive buildings standing empty or stock arriving without sufficient facilities completed.

Remember too that planning and building almost always take longer than you think. There is a chain of people involved including local authorities and suppliers. Any delay in this chain holds up the next stage of development. Few buildings can be planned and erected within the time originally allocated, so forward planning and a realistic time-scale are important to avoid any form of compromise at some later stage.

Chapter 3

THE ENVIRONMENTAL NEEDS OF THE PIG

ALTHOUGH GREAT care in planning a unit or building should always be taken, the most influential feature in the successful operation of any piggery is the achievement of a good standard of environmental conditions required for pig production.

Environmental control frequently attracts insufficient attention compared with other components of a unit such as pen dimensions, choice of materials and the like. In addition many factory-built buildings have suffered by being fitted with temperature control equipment inadequate for the purpose intended or equipped with an air movement system based on unsound principles. These particular problems have tended to bring mechanical ventilation systems into a certain amount of disrepute. It is entirely understandable that a farmer requiring to make a choice of environmental control for pig buildings should first seek to satisfy the requirements of the pig by 'natural' means.

It is often difficult for the farmer to assess adequately the part of the proposed design which deals with the suitable environmental provision for a pig house although this is of vital importance to efficient production. The needs of the pig have been fairly clearly defined by research work in many countries. However a whole combination of factors can combine to defeat the designer's intention.

TEMPERATURE

The main standard by which environment is judged in a piggery is air or house temperature. Although this is not the only important characteristic it is the most important one and all other factors, except possibly light, interact with temperature to produce the level of pig comfort required.

To create suitable temperature conditions for pigs it is necessary to attempt to create a buffer between the circumstances created by nature, added to by the pig itself and compounded by the building in which the animals are placed. Temperature requirements may not be too difficult to define but the

35

presence of other factors such as draughts, the provision or lack of bedding, stocking rates, size of pigs, group sizes, insulation effectiveness and scale of feeding means that temperature itself cannot be taken in isolation. All these factors affect the pig's reaction to its environment, not to mention its health status.

The pig can, to some extent, compensate for environmental shortcomings but it is easy to abuse this. It is most important to realise that young piglets simply *cannot* cope with inadequate temperatures. It is essential to make provision for environmental standards which are known to give positive results rather than starting with a compromise based on increased feeding or adding more bedding to compensate for suspect environmental circumstances. The installation of a new building or the modification of an existing one offers a new opportunity to achieve the correct environment, and this should not be missed.

Producers must also accept the fact that they themselves can markedly affect otherwise sound design and constructional details by faulty operation and by uninformed judgements. For example, ill-fitting doors or slurry channel covers can create extremely powerful draughts which may create microclimatic conditions within certain limited parts of the house. These may not show up on a temperature sensor sited some metres distant. All too frequently, control equipment is badly maintained leading to gradually declining effectiveness. Many operators are not sufficiently well informed to understand that misinterpretation of control equipment can seriously affect the comfort of the piggery.

The provision of the correct temperature is vital. Standards are known but the means of providing and maintaining that environment are many and complex.

The climatic environment required by the pig to maintain a constant body temperature will vary according to the size of pig, structure of the building, feed scale, air speed at pig level and other sources of heat. The figures given in table 3.1 should only be used as a guideline and it is essential that the producer not only monitors the environment frequently but is ready to interpret pig appearance and performance in an informed way. He should be prepared to make adjustments even if this means an absolute temperature fall outside that shown, in order to achieve full performance. The table specifies air temperatures because these are easiest to measure. Allied to this is the fact that *most* of the heat loss from a building occurs in the form of air transfer, such as that created by the ventilation provided or draughts. The producer should give control of the environment greater priority than insulation but it should not be treated in isolation from building structure design.

The Concept of Critical Temperatures

Examples in table 3.1 illustrate the temperature ranges acceptable to the pig dependent on age, feed intake and body condition.

Lower critical temperature (LCT) is the lowest temperature at which heat loss is minimal and efficiency therefore greatest. The aim of building design is to reduce the large difference between the temperature of the pig and its environment (i.e. cold weather conditions produce an increase in heat loss). Below this figure feed has to be provided to maintain body temperature.

Upper critical temperature (UCT) is the upper temperature at which pig performance is affected by, for example, being less comfortable, having a reduced appetite and, in extreme, suffering heat stress.

Thermoneutral zone is the safe temperature range for the class of pig under consideration. The amount of heat loss does not affect full performance efficiency.

Table 3.1 Examples of production circumstances and related temperature requirements

Class of pig	Typical weight range (kg)	Type of housing and management	Thermoneutral zone (°C)*
New born	1.0–1.5	Heated creep needed to provide correct temperature.	28–30
Suckling	1.5–5.0	Insulated, concrete creep floor	24–29
Suckling	1.5–5.0	Totally slatted, metal creep floor	25–30
Sow and litter	Sow only	Pre-farrowing–2 days post farrowing	20–25
,, ,,	Suckling period	Bedded farrowing pens	16–24
,, ,,	,, ,,	Totally slatted farrowing pens	17–25
Service area (Sow/Boar)	120–200 kg	† Sows fed 3–3.5 kg/day in service area	
,, ,,	,,	Individual stalls, insulated, no bedding	17–26
,, ,,	,,	,, ,, totally slatted	18–27
,, ,,	,,	,, ,, part slatted	18–27
,, ,,	,,	,, ,, with bedding	15–25
,, ,,	,,	Group housed, totally slatted	15–27
,, ,,	,,	,, ,, with bedding	10–25
Dry sow	120–180 kg	† Fed 1.8 kg/day individual pens, part slatted	17–26
,, ,,	,,	† Fed 2.2 kg/day ,, ,, ,,	15–21
Dry sow	120–180 kg	† Fed 2.2 kg/day in group bedded system	14–20
,,	180–220 kg	In good body condition, unbedded stalls	14–20
,,	180–220 kg	In good condition, bedded, group system	13–19
Weaners	5–7 kg	Fully slatted on metal floors	29–32
,,	,,	Insulated concrete floor	29–31
,,	,,	Bedded floor	27–30
Growers	7–15 kg	Fully slatted on metal floor, full fed	22–27
,,	,,	Insulated concrete floors–full fed	19–26
,,	,,	Bedded floors—full fed	18–25
,,	15–30 kg	Fully slatted, metal floors—full fed	17–26
,,	,,	Insulated concrete floors—full fed	14–24
,,	,,	Bedded floors—full fed	13–23
Finishers	30 kg +	Fully slatted concrete floors—hopper fed	15–27
,,	,,	Fully slatted concrete floors—restrict fed	17·27
,,	,,	Part slatted floor—generous feed level	17–25
,,	,,	Bedded floor—hopper fed	11–22

* Safe air temperature (assumes satisfactory air speed)
† Assumes pigs fed in excess of maintenance needs by virtue of diet quality—degree shown by feed scale.

It must be stressed that the temperature ranges quoted in these examples are governed by the housing/management structure and if taken out of context, the LCT/UCT would alter.

Total heat production by the pig increases with both body weight and plane of nutrition. This is in direct contrast to the needs of the pig which decline with increases in bodyweight and feed scale. This disparity between nature's provisions and the pig's needs creates problems for the building designers. In sows, critical temperatures will be higher for those in poor body condition, on lower planes of nutrition and in the earlier stages of pregnancy.

Factors which Influence the Rate of Heat Loss

Heat loss is governed by conduction, convection, radiation and evaporation. The rate of heat loss can be influenced by both the circumstances under which the pig is kept and the way it is managed.

Nutrition. Diet quality and quantity both influence pig output. When the energy intake exceeds maintenance requirements it produces increasingly more heat output. This is indicated in table 3.1. The corollary of this is that higher feed scales reduce LCT for any size of pig.

Size of Pig. The larger the animal the greater its heat output and (as shown) the lower its LCT. Conversely the small pig has a greater tolerance of high temperature.

Penning Arrangements. Where pigs are penned in groups they lose heat less rapidly. Lying against each other, they expose a smaller area of their bodies from which heat is lost.

Structure. Table 3.1 demonstrates the influence of floor type on heat loss. This will also apply to the degree of insulation used in other parts of the structure, particularly the roof.

Air Movement. The greatest heat loss from a building is by way of the ventilation system. Air movement over a pig or group of pigs has a cooling effect and this will be affected by skin wetness, presence or absence of bedding, the temperature of the air and its actual speed.

Assessing House Temperature

Measuring house temperature is not particularly easy due to localised variations existing within a house. Mechanical ventilation is often unreliable due to the position of the sensor, accuracy of the controller and the precision of design and operation of the house. The dial setting on the control panel should not be taken to be the actual temperature. Sophisticated equipment does exist for precise temperature measurement but in practice a producer will be guided by pig behaviour and performance in deciding the suitability of the temperature control. Growth efficiency, feed intake, scours and even mortalities in the extreme cases, soon indicate whether a massive discrepancy between desired and actual temperature exists. Other common signs are fouling of the pen floor, pigs huddling or, conversely, panting as pigs attempt to alleviate their discomfort.

The two checklists below give general guidelines on eradicating obvious defects within an existing building where temperature discrepancy from the thermoneutral zone is observed or measured. In any case re-calibration and checking of control equipment is always worthwhile at routine intervals.

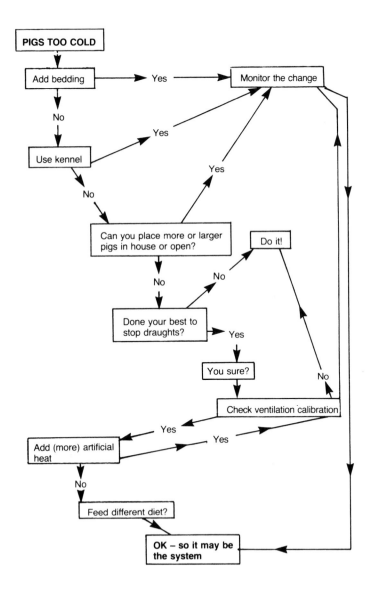

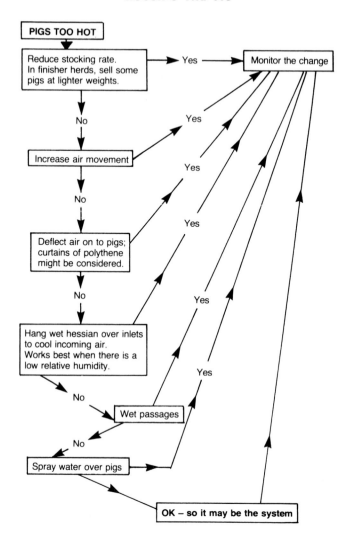

Regular careful cleaning of ventilation ports and other parts of the equipment is also important.

The probable avenues of heat loss or uncontrolled gain have been underlined. Differences in approach to this problem will however become apparent when new designs are proposed. The purchaser will need to exercise judgement upon actual design proposals as well as choice of materials. Efforts to inspect buildings currently in use of the type projected is well worthwhile. However, this approach still creates problems. Care should therefore be taken in drawing conclusions as to the value of a particular layout on the basis of a single visit.

- Circumstances can change with different seasons of the year.
- Change in stocking rates and size of pig can make a great impact.
- The type of feeding practice used may not be appropriate to the projected building.
- Pig health can distort the perceived effectiveness of a building.
- Some designs, particularly those where manual adjustments are necessary, are very dependent upon the skill of the individual operator. Careful appraisal of the availability of such inputs must be honestly made.

Providing these considerations are openly applied there is some merit in inspecting designs in use in order to evaluate the proposals made. The choice of floor system and use of bedding will highlight the relative importance of thermal insulation yet for *any* structure the accuracy with which the house temperature can be buffered against the prevailing ambient conditions is important.

Insulation

Good insulation enables the sensible heat production from the pigs to be harnessed in making a major contribution towards the total temperature needs. It also minimises the influence of solar gain upon the higher temperature range. The method of evaluating the amount of insulation required—or that which can be justified—is not easy and a method of calculating the merits of varying materials or insulation proposals requires a basic equation to enable comparisons to be made.

The measurement used to assess structural heat loss is the 'U' value. Tables exist to allow various materials to be compared. The 'U' value gives the amount of heat which will be lost through 1 square metre of the building surface when the difference between the house and the outside temperature is 1°C (the measurement is taken in watts and is expressed as $W/m^2°C$). Therefore, the lower the figure the better the insulation value. As a guideline a reasonable standard of insulation is reckoned to be: 0.6 $W/m^2°C$.

A lower (i.e. better) level may be considered in weaner and farrowing houses. It is worthwhile paying attention to the insulation of doors and windows to avoid additional heat loss from condensation.

Insulating Materials

These may possess various properties such as physical strength, moisture, flame and vermin tolerance as well as different appearances. Blocks or flooring materials can form part of the structure whereas flexible quilts, foams etc. can only be used as insulation barriers. Some materials are moisture tolerant, but if not, the material used must be properly sealed to prevent any moisture damage which would reduce the value of the insulation. Unfortunately most insulation materials are prone to vermin attack. Proper sealing can help in preventing such damage but constant vigilance and regular control measures for rodent and bird damage are essential.

Spray-on urethane foams are becoming more widely used. They have high insulation value, and can be used for sealing gaps and cracks without the need for any structural supports or extra sealing. These qualities make them particularly useful for building conversion work. Careful, professional application is necessary and only reputable suppliers should be used.

A quick guide to the relative properties of the more common insulation materials is shown in table 3.2 assuming no added benefit from air spaces and that a single covering of asbestos is used.

Table 3.2   The relative properties of insulation materials

Thickness of insulation (mm)	Glass fibre or mineral wool	Expanded polystyrene	Extruded polystyrene	Polyurethane spray or panel
25	1.24	1.09	0.96	0.79
50	0.7	0.61	0.53	0.55
75	0.69	0.43	0.36	0.47
100	0.37	0.32	0.28	0.29
125	0.30	0.27	0.22	0.18
150	0.25	0.22	0.19	0.15

The broken line indicates the level above which the 'U' value may be too high

Sealing proposals for doors, windows, gates, slurry chambers and ventilation ports in any new design should be closely studied and an assessment made of the standard of draught elimination achieved. Every effort should be made to avoid uncontrolled air movements in the building; they are particularly dangerous due to their very unpredictability.

VENTILATION

This is the major pathway of heat loss from a building. Ventilation has to fulfil several needs:

- To control air temperature.
- To control relative humidity.
- To maintain a satisfactory degree of comfort for stock and operator, including toxic gas levels.
- To quantify the speed at which air passes over the pigs.

Due to the difficulty in measuring relative humidity, the air temperature is normally used to determine the ventilation rate required. However this may conflict with some of the requirements listed above because of humidity and odour levels rising in cold weather when it is necessary to ensure that the optimum temperature is maintained and air is not passing too rapidly over the pigs. The use of air temperature measurement may well be appropriate, therefore, to keep a building within the Critical Temperature range.

The suitability of a ventilation system to provide the desired temperature range may be judged in two parts: low ambient temperatures and high ambient temperatures. Workers at the Scottish Farm Buildings Investigation Unit recommend that the likely temperature range in the locality for which the building is designated should be taken into consideration. Readings from the local meteorological station will establish the likelihood of a temperature being above or below a particular range and this can be taken into account when deciding the required insulation and ventilation levels. Using such an approach a producer can decide whether the design parameters proposed for the site will suit the location. In particular this approach is well suited to an assessment of insulation standards to be adopted and upper ventilation rates required for a building.

The provision of minimum ventilation rates poses a substantial problem in pig housing. There will always be conflict between the standard of temperature required and the humidity/odour levels present. Maintaining a satisfactory minimum temperature using supplementary heating is costly. Some environmental scientists propose that carbon dioxide (CO_2) levels be used in determining the *low ambient temperature conditions* for a specific design criteria. The standards in table 3.3 provide guidelines on ventilation for various pig categories and is based on the calculated CO_2 levels predicted for various classes of stock at the lower temperature range and given normal feeding practices in the production stages shown. A maximum CO_2 concentration of 0.3 per cent is generally recommended although adult stock may tolerate a slightly higher level (0.5 per cent).

Table 3.3 shows the ventilation rate recommended for higher temperature conditions where a temperature lift (caused by pig heat output over air temperature) of 3°C or 4°C is assumed. It will be noted that a considerable difference in ventilation capacity is needed to give an additional 1°C of control under typical circumstances and again it is important to remember that smaller pigs have a greater tolerance of the UCT than older categories.

In using table 3.3 to assess the likely efficacy of a ventilation system it is advisable to note the following points:

- Precise house stocking standards and the probable local climatic range should be calculated for each house.
- The use of *average* body weight figures may lead to unacceptable standards for small pigs in cold conditions and also larger pigs in warmer weather. There may be a distinct variation in the bodyweight of individual pigs over a relatively few weeks in their life, which will affect heat loss calculation and environmental needs. Therefore the cost of incorporating more ventilation or supplementary heat would have to be considered, particularly if an all-in/all-out policy of operation is used.

A widely used system of ventilation is variable speed fan control. These fans are designed to operate over a range of their total speed/output capacity,

Table 3.3 Ventilation Guidelines

| | Ventilation rate in $m^3/h/pig$ housed | | | | Ratio between min/max rate at 75% pen occupancy | |
| | Minimum CO_2 concentration level | | Maximum Acceptable temperature lift over ambient | | | |
Category of pig	0.3%	0.5%	3°C	4°C	Continuous stocking	All-in/All-out
Service	13	8	190	140	32	—
In-pig sows	13	8	150	110	25	—
Farrowing House	13	8	480	350	47	80
Weaners (ave 10 kg)	1.3	0.7	38	28	23	72
Grower (ave 25 kg)	3.2	—	66	49	15	28
Finisher (ave 62.5 kg)	6.6	—	135	100	14	27

(From ADAS Booklet, no. 2410)

speeding up and slowing down depending upon the relationship between the temperature within the house and the desired temperature.

- However, the ratio between minimum and maximum ventilation rates given exemplifies the weakness of fan ventilation systems which rely upon the minimum ventilation rate being a percentage of the maximum speed. Many of the figures shown are outside the performance range achievable with a single fan. To avoid over-ventilation in the lower temperature range it may be necessary and desirable to have a two-stage system whereby Circuit 1 controls temperature up to the desired level and Circuit 2 acts as a booster system to increase the ventilation rate whenever the set temperature is exceeded.

Another important aspect to consider when judging ventilation proposals is the amount of protection provided in the event of system or power failure. Protection of the stock may be arranged by installing panels held in place by magnetic catches which fall away when the power fails. This allows some natural air flow. More sophisticated systems may automatically switch to standby generation equipment. Certain sensitive categories of pigs, particularly the young, may also require back-up protection from fan or control panel breakdown or heater failure. This is often achieved by the use of an upper and lower temperature sensor connected to an alarm system which will give warning of malfunction.

The Influence of Air Speed

The pig is very sensitive to the influence of air speed because of its poor body insulation. This problem can be affected in the following ways:

- Where temperature is below the lower limits of the comfort zone—see table 3.4.
- Pig body condition is less than ideal.
- Nutritional inputs, especially energy, are for whatever reason low.
- Pigs are penned individually—see table 3.4 below.

Air movement in excess of 0.2 m/sec. across a pig will generally create discomfort but it should be borne in mind that a lower speed will be necessary where the above four conditions are not met.

Table 3.4   Influence upon Lower Critical Temperature of different parameters for Pigs in range 40–100 kg

Weight of pig (kg)	40 --- 100					
Group size in pen	4 ------ 20 --- 20					
Air speed (m/s)	0.5 -------------------- 0.2 --------------------------- 0.2					
Type of floor	Concrete ---------------------- straw --------------------					
Feed scale (× maintenance)	2.5 --- 3 --------- 3					
	20°C	*17°C*	*14°C*	*11°C*	*8°C*	*5°C*

Lower Critical Temperature

If excessive air speed was imposed upon suspect temperature conditions and low group size it is obvious that piglet performance would be even lower than that generated by operating at below the Lower Critical Temperature as table 3.5 indicates.

Table 3.5   Effect of group size on pigs' heat loss

Piglets per group (5 kg bodyweight)	Effect on heat losses at two different temperatures
1	Lost 47% more heat at 20°C than at 30°C
4	Lost 27% more heat at 20°C than at 30°C
6	Lost 11% more heat at 20°C than at 30°C

Cubic Air Space

To increase the provision of air space in a building is expensive because it requires a larger structure. There are practical limits however to the amount by which the cubic space of a house may be reduced.

The larger the volume of air within a building the bigger the buffer created against high air speeds at pig level. Providing this space is not excessive, there should be no difficulty in maintaining a constant temperature. In buildings with a low cubic air space, ventilation control at low air speeds is more difficult because the rate of air exchange at any given fan speed/ventilation rate will be greater than in a building with greater volume.

Where pre-heat chambers are used for the warming of air entering a building a ratio of 1:10 is normally necessary to generate a practically operable cross-ventilation system.

Table 3.6   Guidelines for cubic air space provision

Category of pig	Cubic area per animal (m^3)
In-pig sows in stalls/tethers	4.5–5.75
Farrowing house	4.5–12
Weaners (5–10 kg)	0.4–0.8
Growers (10–20 kg)	0.8–1.3
Finishers (20–90 kg)	1.3–2.3

Provision of Artificial Heating

This would only apply to adversely low ambient conditions, the likelihood of which would be obtained from a study of local meteorological records. If the heat output from the anticipated size and number of pigs in the building is plotted against heat losses through the structure and via ventilation and draughts the deficit can be calculated. Many advisory organisations now possess computer programmes which include parameters to account for insulation standards, ventilation rate, quality of draught-proofing, feeding regime to be used, size and weight range of pigs, numbers per pen, etc.

Organisations capable of providing a service for such calculations in the United Kingdom include ADAS, some university and college departments, Scottish Farm Buildings Investigation Unit, Farm Electric Centre, National Institute of Agricultural Engineering and various commercial companies.

In general farmers should be aware that:

- Badly controlled ventilation is likely to be the major cause of heat loss.
- Draughts must be excluded.
- Insulation standards recommended must be adhered to.
- Major changes in feed policy and stocking/penning methods will alter the requirements of stock and performance of the house.
- It is always worthwhile employing the services of an advisory agency to evaluate the specified components of house temperature design before accepting any proposals.

Principles of Air Movement

Even though the quantity of air moved may be correct, the actual environment around a group of pigs will depend on how the air reaches them. Although special situations may give variance, it is usual that the air flow pattern within a piggery consists of large rotary motions. The actual pattern of air will depend upon the following main points:

The design of the air inlet. This has a dominating influence. If incoming air is a different temperature (and therefore has different movement properties) it can radically influence the air flow pattern within the building. Conversely, outlets have only a small localised impact on air movement patterns because exhaust air reaches outlet points from all directions. Air inlets can be designed to create a jet effect which generates a predetermined pattern to increase predictability and, therefore, animal comfort. Research shows that constant air speeds of 5 m/sec give a predictable inlet and movement pattern. Careful design features are demanded within a piggery where ambient air enters a building with a high velocity such as 5 m/sec. It should not pass over the pig lying area at pig level greater than 0.2 m/sec. This requires the air inlet size of an extraction system to be automatically adjusted in concert with the fan speed.

Position of obstacles. Incoming or relatively fast moving air will be disturbed by quite small obstacles such as a 5 mm diameter electric cable.

The position of the stock. Convection currents can create their own air pattern where air speed is low over the pigs, i.e. the inlet is too large relative to the speed of the air passing through it. In fact, some ventilation systems create an opportunity for convection to become the dominant feature. The main problem with this is that it is unpredictable. It is also difficult to harness and use as a part of the ventilation pattern or system.

Fan ventilation systems which have adjustments down to very low speeds are prone to wind interference. This not only disturbs the airflow pattern but may create damage to fan motors. It is possible to overcome this problem by carefully baffling the air inlets and using higher fan speeds (i.e. smaller fans moving relatively faster for low temperature control).

Whichever ventilation system is used, it is important to maintain a degree of control over the direction of the ambient air entering the building.

Which Ventilation System?

The majority of pigs housed in intensive units are still subject to ventilation designs which do not stand comparison with the guidelines under discussion. Much of this is due to financial compromise, particularly where the capital cost of the ventilation system has a large influence on investment decisions. Mechanical ventilation has become the target for considerable criticism as a result of poor pig performance and comfort. It is not this kind of ventilation in itself which is wrong but the total design of the system or some of its component parts, or perhaps the operation of the house.

There are three major ventilation systems:
—Natural ventilation, usually with some manual adjustment.
—Automatically controlled natural ventilation.
—Fan ventilation.

Natural Ventilation
There are two forces which may be utilised to effect ventilation of a pig building: wind effect and convection or stack effect.

The *wind effect* is simply that wind blowing against one side of a building creates suction pressure, thus admitting and exhausting air on opposite sides.

In conditions when the air is still, or moving slowly, the heat from the pigs will rise, the *stack effect*. If this air can escape at a high level it will be replaced by cold air entering the building. The greater the height that this air has to rise the greater the force needed to draw in the fresh air. If the warm air cools too much as it rises the stack effect will diminish.

The largest influence on the success of a natural ventilation system using stack effect is the ratio of air outlet to inlet. The recommended ratio is that the inlet air should be *twice* that of the outlet. This arrangement usually creates a satisfactory stack effect but it may cause draughts at pig level necessitating the closure of some of the inlets. It has been shown that, given the known ventilation rate for the pigs, their heat output and the height between inlet and outlet ports, the air capacity can be calculated.

Natural ventilation is controlled by the manual adjustment of flaps, normally situated at the ridge or the eaves. A refinement of adjustable lids, flaps and pophole shutters may also be added because it is normal for some form of internal kennelling to be incorporated in such a building. In the main building itself some control of ventilation at the eaves is important. Otherwise in cold conditions heat loss could be excessive and cold air could enter the ridge in an uncontrolled fashion, increasing the chilling effect.

Eaves adjustment offers greater accessibility than ridge adjustment and is essential in a pig building. A Dutch barn type structure of over 18 m (60 ft) in width will be almost entirely dependent upon stack effect rather than wind effect to achieve air movement. If space boarding is used at the eaves of barn-type structures there must be some means of enclosing part of it in windy conditions. It may also be necessary to be able to open these during still, warm periods if polypropylene mesh is used. Experience suggests that to keep rain and snow out with an open ridge some form of covered inlet is probably required, with an overhang twice the size of the ridge gap. This or a baffled chimney are preferable to upturned flashing.

One of the most successful naturally ventilated buildings is the low monopitch design where building length is twice the width and the front height is twice the rear height with floor to ceiling divisions between each pen. This is a case where the building for growing and finishing pigs justifies the cost saving on electric controls. Natural ventilation demands regular input by the operator as it is difficult to control in the extremes of temperature. Young pigs are temperature sensitive and this type of housing is unsuitable for them.

Automatically Controlled Natural Ventilation (ACNV)
The system consists of adjustable flaps or curtains in the eaves wall of the building with mechanical adjustment, temperature sensor and a control panel. The flaps are adjusted if the temperature varies from the pre-set level by automatic checking of the temperature at regular intervals. The amount of adjustment depends on the frequency and length of time during which the temperature is checked, and the duration for which power is provided to the motor to adjust the flaps. Flaps are normally designed to be centre-pivoted as these require less mechanical effort for adjustment. It is important to the successful operation of the system to ensure efficient flap design. Light, insulated shutters have much to commend them as they will resist condensation and may be moved with little mechanical effort. Standard calculations relating to this system based upon the size and number of pigs to be housed and the locality, have been prepared for general use (see references).

It is a low-cost system and quiet to operate and is capable of good temperature control for a range of pig categories. It can be made virtually fail-safe by ensuring that the flaps never quite close. However, precise temperature control, for early weaned pigs for example, is not possible. The direction of air will vary which in turn will possibly produce variation in pig lying and dunging habits. Furthermore there are limits to the building width which may be successfully ventilated using this system; 9 m (30 ft) span is close to the upper limit possible. There may be a case, as often seen in the

United States, for ACNV to be part of a hybrid system which includes some ridge-mounted fans fitted with backdraught shutters operating only when high temperature levels prevail.

Fan Ventilation
The main drawback of this kind of ventilation is high running costs, and previous experience of its use in the pig industry has been largely unsatisfactory. Noise, dust and odour expulsion, particularly in areas close to residential zones, have also been limiting factors. Fan ventilation systems may be categorised as follows:

Extraction systems will only work well where there is a degree of control over the inlet:extraction rate. This can be achieved with variable speed fans but works better where a number of fans are switched off and protected by back draught shutters during periods of lower temperature. Too many of these systems are over-ridden by wind when the fans are operated at low speeds and this also leads to a degree of unpredictability which creates variation in pigs' behavioural patterns.

Pressurised systems using mineral wool or some other dispersal medium as the base of a loft through which air is forced does give very low air speeds over the stock. However at slow fan speeds, convection currents tend to take over and create unpredictable venting which can lead to a build-up of humidity and dust. These systems must be fitted with fail-safe mechanisms.

Pressurised systems using rigid dispersal baffle are also used to give low air speeds over pigs. As in all pressurised systems unplanned gaps become outlets (rather than inlets as in pressurised systems) so there are benefits in preventing draughts around doors, etc. Outlets should be baffled against wind effect.

Recirculation ducts with air distributed from suspended perforated polythene ducts may be used to give an opportunity for some re-use of air within the building. Proprietary ridge-mounted units also achieve this result, but these work best within larger buildings. The recirculation duct has to be gable-mounted so may be confined to part of a system which also includes some extraction units for warmer weather conditions. One disadvantage of these layouts is that they tend to encourage the build-up of dust within the building which may also increase humidity and odour levels.

Cross-flow systems work best where air enters the pig area via a plenum chamber and is suited to the room principle of housing commonly used for earlier-weaned piglets. Care has to be taken to sub-divide the plenum area for each section to prevent air being drawn from room to room when differing fan speeds are in use.

Properly designed fan ventilation offers the greatest opportunity for precise control of the environment for all classes of pigs. Used in conjunction with a heater interlock facility it is capable of providing accurate and consistent control for the youngest stock.

Which Controller?
A variety of control systems are available for various ventilation systems.

Proportional controller. This is used to provide variable speed of fans within the range of fan speeds from 5–10 per cent up to 100 per cent of their capacity. These should have a short step (usually 1°C either side of the set temperature) where they change from minimum to maximum speed. Poor control of house temperature will result where the controller fails to operate over the short step or work at the upper and lower limits set.

Stepped controller. This controller switches fans on and off in sequence according to the variation recorded between the house and the set temperatures. The fans are run at full speed and are consequently less susceptible to stalling at low speed or when wind is high. When they are not running it is essential that the fans are each protected by a backdraught shutter. It is often so arranged that one small fan runs continuously to supply minimum ventilation requirements.

Timed on/off controller. This is often used in conjunction with the stepped control system to give minimum ventilation rates with frequency of running based on operator experience.

On/off controller. This is a simple thermostat-operated system which switches fans according to variation from pre-set temperature levels.

Interlock controller. This controller operates a system where whole-house heating and ventilation is employed for more economic energy usage. Fans are only allowed to increase from lower speeds when the heaters are off. Some models can also switch heaters on in stages to provide even closer control.

Computer controller. Sophisticated programmable panels are now available which adjust to pre-set levels of growth, such as in a 'room principle' grower and finisher house where pigs grow in one section of the house over a number of weeks.

Which Fan Size?
Propeller fans are normally used in pig houses. These are usually mounted on a diaphragm plate if positioned in ducts or on a bellmouth mounting if wall fixed. Care must be taken to match fans by size when working within the same area.

Table 3.7 shows typical performance of fans and is prepared by the National Institute of Agricultural Engineering.

The check list below the table on page 52 proves that the *possibility* for effective control exists if correctly designed, installed and operated!

Table 3.7   Performance of propeller fans at 5 mm water gauge in different fittings

Fan diameter (mm)	Speed (rev/min)	Mounting arrangement	Volume (m³/sec)	Area (m²)	Mean air speed (m/sec)
315	1380	Bellmouth	0.38	0.078	4.9
		Diaphragm	0.31		4.0
400	1360	Bellmouth	1.03	0.126	8.2
		Diaphragm	0.89		7.1
450	1350	Bellmouth	1.6	0.16	10.0
		Diaphragm	1.4		8.8
500	900	Bellmouth	1.0	0.20	5.0
		Diaphragm	0.9		4.5
630	700	Ring	1.6	0.32	5.0
		Diaphragm	1.5	0.33	4.5

Why don't ventilation systems work?

The Designer
Too much compromise to achieve a 'sale'.
Too much compromise in assessing pig requirements.
Lack of know-how.
Over-complicated design.

The Installation Engineer
Poor understanding of designer intentions.
Poor calibration.
Lack of operator instructions.
Poor follow-up/maintenance.

The Operator
Failure to use building for purposes it was designed.
Poor understanding of the ventilation system installed.
Poor equipment maintenance.
Failure to check calibration of equipment.

Summary

Whilst combinations of ventilation systems and control methods can work within a variety of house designs, *all* ventilation systems work better where:

- Pigs of a narrow weight range share a given air space.
- Wide control is possible to mask the extremes of climate and stocking rates likely to be encountered.
- Buildings are not too wide unless air ducts are used.
- Air deflection is used to counter discrepancies in ambient temperature.

- Distinct temperature zones are formed by kennel construction or good air deflection to encourage clean lying habits of the stock.

Because of the inter-relationship between all these factors it is impossible to eliminate any combination as being unsuitable, yet as the checklist shows, disappointment may ensue if designer, installer or operator fail to contribute their skills properly.

OTHER ENVIRONMENTAL CONSIDERATIONS

Humidity

Pigs cope well with a wide range of relative humidities (RH) except at the extreme of temperature ranges. For example when temperatures are high and RH is high, little evaporative cooling can take place; high RH at low temperature will increase the chilling effect. Pigs perform well under a RH range of 60–85 per cent. The humidity level is important more for showing the effectiveness of the ventilation system rather than for its direct importance to the stock.

Noxious Gases

Under normal piggery conditions gases should not constitute a major problem. However when slurry is agitated, levels of hydrogen sulphide, carbon dioxide and ammonia may rise rapidly and care with ventilation must be taken at such times. As already mentioned a figure of 0.3 per cent for CO_2 is considered acceptable and this is undetectable by smell. Ammonia becomes apparent at a concentration of 5 ppm and could be a serious problem to both operator and pigs at figures above 20 ppm. Hydrogen sulphide can be smelt at 0.1 ppm and would indicate inadequate venting when slurry was moved.

Pig Health

It is fair to say that all systems of housing work well where pigs have no serious clinical health problems. However, there is a further truism that subclinical health problems will ruthlessly expose deficiencies in temperature control particularly in respect of respiratory conditions.

Floor Cleanliness

The cleanliness of lying area floors is very important. Frequently pigs 'turn dirty' when temperatures reach either extreme but more particularly when hot. Inadequate ventilation capacity and poor directional flow can contribute to this problem as can the ratio of dunging to lying floor area (1:1 ratio is best), in penning which is not fully slatted. Floor feeding and a reduction in stocking rates will also help to keep floors clean. The deflection of incoming air by the use of polythene drapes during hot weather can be beneficial and also the wetting of passage floors to achieve cooling by evaporation. Where there is a low RH at high ambient temperatures, i.e. in parts of North America and Central Europe, evaporative coolers can markedly reduce house temperatures and increase pig comfort.

Waste Heat Recovery

Attempts are now being made to harness the heat from stale air to raise the temperature of incoming air whilst keeping them separate. In addition to this method, a heat pump system may also be considered whereby the air is condensed to heat water, which in turn may be used to heat other areas of the unit where there is a temperature deficit. Clearly, the economic feasibility and justification for such systems depend on the exhaust heat losses available which are proportional to the volume-of-flow rates and the difference between the outside and in-house temperatures. Capital costs of the equipment provide a major obstacle to their widespread adoption at present. Rising feed and fossil fuel costs as well as an increasing need for greater operator comfort are likely to generate active interest in these systems.

Standby Generation

The need for back up power generation increases as environmental control and feeding become increasingly mechanised. Risks to animal welfare are greater on farms with a history of supply problems. The choice of standby capacity will depend on the likely degree of risk and the funds available for this purpose. Bare essentials only could therefore be safeguarded or a serious attempt made to operate the unit normally on the standby mode. About 23 kW per 100 sows in a breeding/feeding unit may be required for whole-unit heating, ventilation, feeding and lighting.

Lighting

Although human eyes have a considerable tolerance to a wide range of illumination conditions, efficiency of operation and safety will be improved where adequate lighting is provided. There may even be a level below which pig performance may be depressed. For normal inside operations a light intensity of at least 50 lux is required. Prolonged periods of lighting within a house at this level may tend to encourage greater aggression in the pigs. The normal lighting types recommended for use in piggeries are medium to low such as dispersive reflectors or 'Coolican' reflectors. Fluorescent lights, preferably in enclosed casings, give greater dust/moisture tolerance. As well as satisfying health and safety requirements BS 5502 also makes recommendations. The Farm Electric Centre produces tables which allow designers to produce satisfactory lighting patterns. One of these is reproduced as table 3.8, to facilitate a check on the suitability of a lighting plan recommended by a buildings supplier.

Table 3.8   To provide 50 lux illumination using totally enclosed white fluorescent lamps

Building span (m)	Eaves height (m)	Number and wattage of lamps	Rows of lamps	Spacing (m) across span × along span	
6.0	3.0	1/65	1	central	4.0
	4.8	1/65	2	3.0	6.0
	5.4	2/65	2	3.0	6.0
7.2	3.0	1/65	2	3.6	4.8
	4.8	2/65	2	3.6	7.2
	5.4	2/65	2	3.6	4.8
9.6	3.0	1/65	2	4.8	3.8
	4.8	2/65	2	4.8	4.8
	5.4	2/65	2	4.8	4.8
12.0	3.0	1/65	3	4.0	4.8
	4.8	2/65	2	6.0	8.0
	5.4	2/65	2	6.0	8.0
14.4	3.0	1/65	3	4.8	4.8
	4.8	2/65	3	4.8	7.2
	5.4	2/65	3	4.8	7.2

Chapter 4

STRUCTURAL CONSIDERATIONS

EXISTING PIG buildings demonstrate the widest use of materials of almost any group of structures. The willingness of pig producers to try out various types of partitioning or flooring has led to many materials receiving reputations—both good and bad—which are unjustified. This chapter attempts to illustrate the way in which a choice from the ever-growing list of materials might be satisfactorily made.

Materials used in any part of the structure must be chosen to ensure that they satisfy the following necessary requirements: durability, ease of cleaning, comfort, slip resistance (where relevant), weather resistance, aesthetic acceptability, fire resistance, dust tolerance, thermal insulation, resistance to slurry/manure corrosion and acoustic properties. All such considerations are dealt with in considerable detail in the BS 5502 (see Chapter 14).

Experience has shown that many materials have come to be thought unsuitable for use in piggeries, not because they have inadequate properties but because they have been fixed inadequately. Any material selected for use must be fixed and supported according to the relevant manufacturer's recommendations or British Standards. The recommendations will ensure that adequate length of life and protection are achieved. Accuracy of construction is a major consideration with main structural components and close adherence to recommended tolerance levels for deviation should be adhered to (BS 5502 section 3.9:1978).

The major factors determining the size of a building will be the needs of the pig and the operator, allowing for ease of pig movement, feeding, dung removal and inspection. Whilst these considerations should always be given priority it is also relevant to consider choice of bay sizes and the dimensions of joints and profiles so that the maximum interchangeability of materials and the most cost-effective choices can be made.

Siting can have a considerable impact upon structural design. The effect of high winds, snow and rainfall vary with location. Furthermore, the exact positioning of a building will affect its relative exposure. If the building is sited in the lee of a hill or surrounded by natural windbreaks or other buildings, some design factors will be different from those in a building sited in a more open position at the top of a hill. The pig producer need not concern himself with such details of construction engineering, but should ensure that the builder or supplier is aware of the on-site circumstances and that any building modifications take into account the site conditions in relation to the materials to be used and the method of fixing. It would suffice to specify that the material and method of fixing employed comply with the relevant British Standard. This will take into account snow, wind and dynamic loadings. Many prefabricated buildings have been approved on the basis of construction methods for use under most conditions.

The Roof

In almost all pig buildings the roof must be weatherproof and of suitable insulation standard. For all practical purposes corrugated roof sheets will be used and there are four main types: fibre-cement, galvanised steel, aluminium and plastic. When profiled sheets are fitted attention must be given to rodent-, bird- and weather-proofing of the terminating point of the corrugations.

Fibre-cement is generally acceptable because its colour changes to a neutral mid-grey colour over time if it is untreated. Epoxy-resin coatings can be applied to new roofs of this material to prevent fungal development and to control the extent of the colour change. Fibre-cement roofing has the great advantage of being essentially maintenance free if fixed according to recommendations. The handling weight and the general reluctance to utilise asbestos products due to their potential health-damaging properties, mitigate against asbestos cement, but fibre-cement alternatives are now available. Appropriate soffit and barge boards should be used to weather-proof the ends of the building.

Galvanised steel comes in two forms: high-strength steel with sheets of 4.5 mm thickness and ordinary quality steel with sheets of a minimum 5 mm thickness. These can be produced in single length sheets to suit the roof pitch and length, and can be factory-tinted to provide an acceptable colour. They are lighter to handle than the same area of asbestos.

Aluminium alloy panels are also light and of variable length to suit the individual structure. The extreme light weight of such sheets demands very careful on-site handling to prevent stress caused by wind or accident. Their highly reflective nature has an added solar resistance but also precludes their merging with the locale, though coloured sheets are available.

Plastic sheets have the advantage of being extremely convenient and easy to cut, handle and fix, but the generous supports needed must be allowed for when comparing costs.

Building Structure

The trend in recent years has been towards factory-built piggeries using an all timber framework including roof support uprights as well as purlins. Concrete and steel uprights are more widely used for wide, portal-framed structures where stock are kept in kennelled and tractor-cleaned penning below. The move towards prefabrication of buildings and the provision of portal-framed structures has led to relatively fewer brick- or block-built structures being used as load-bearing and roof-bearing structures in their own right.

Timber-framed structures have the advantage of high flexibility and low weight to size ratio. Their widespread use in pig units is due to their convenience, although they need careful treatment to ensure long life. Steel-framed buildings provide extremely strong structures but need occasional maintenance to prevent corrosion, and unless treated will not withstand continued contact with piggery wastes. Concrete frameworks have an extremely low maintenance requirement but lack flexibility, and are difficult to modify.

It is less convenient to build from brick and block work. Construction of load-bearing walls takes longer than the erection of a prefabricated framing but, constructed with care, they have many favourable attributes. The use of the appropriate mortar mix, expansion joints and piers or cross walling means they create their own load-bearing framing as well as providing walling. They are highly resistant to corrosion, fire, water and vermin and, if of cavity construction or thermal blockwork, can be self-insulating. If sealed and coated they can also be easily cleaned and apart from attention to re-pointing and painting maintenance should be low.

The Walls

The use of suitable block/brickwork has been mentioned and it is in this area where a greater variety of materials used may be found. Timber, asbestos, glass-reinforced cement, recycled plastic and various combination and composite panellings are widely used. Ease of cutting and fixing has made many types of wall cladding materials popular. In addition, once fixed they may need no further treatment to make them suitable for pig access and easy cleaning. Recent developments have seen the utilisation of panels of various materials on timber studding, factory-filled with insulating foam under pressure to give walling of tensile and impact strength. However, the choice of cladding material, given the difference between internal and external temperatures and its potential distorting effect on the cladding and the timber framing, has yet to be assessed by long-term usage. In addition to these general points buildings of *all* materials require brick and block foundation works below ground. Relocatable structures require a suitable hard standing.

Brick and blocks are likely to be used in the construction of part of almost all buildings, particularly for foundations, slurry tanks, etc. This highlights their merits of strength and corrosion tolerance. Modern additives help to ensure that brick and blockwork are made non-porous, and brush-on coating materials not only improve this impermeability but can make surfaces easy to clean. Cavity walls, particularly when filled with insulant, hollow or light-weight aggregate blocks with other insulating properties can provide an economic thermal resistant structure.

Reinforced asbestos-cement and fibre-cement sheets are very strong, low in maintenance requirements, fire resistant and easy to clean. As a pen partition material they have an added advantage of low support cost, needing support only at the extremes of the sheet size for all classes of pig. It is the potential health hazard which makes asbestos-cement less popular during the cutting and fixing operations.

Glass-reinforced and cellulose/mineral-strengthened cement sheets possess many of the properties of asbestos-cement panelling with the added merits of being easier to cut and asbestos free.

Timber cladding in many forms is also widely used for walling and partitioning. Plywood, laminated chipboards and hardboards are safe to cut and suitable for many purposes. Most have a surface which is easily cleaned but have reduced pest and flame-spread resistance compared to some other materials. Even with the closer supports specified by manufacturers (which add to material cost and labour), many boards are not as durable as other materials. Many are now available in natural colour tinting which contributes to a satisfactory building appearance.

Recycled plastic is also a safe, easily worked material with good cleaning and low maintenance properties. It is, however, very prone to warping under normal farm conditions and needs fixing at 75 mm centres along its periphery to reduce this effect. It is heavy for the area covered which must be borne in mind if gates and removable partitions are being considered.

Integral sandwich or foam-filled insulation is capable of providing added structural strength and good insulation to composite panelling but the extent of this will depend upon the characteristics of the cladding. Experience suggests that a similar material should be used on each side of the studding so that there is a compatible temperature response to minimise differential shrinkage of the panels. There may also be some difficulty with sealing and securing such panels to ensure moisture rejection and rodent-proofing.

Corrugated steel sheeting is widely used for external cladding of buildings. However it has lower impact strength and is relatively difficult to seal at its longitudinal ends, so it is less popular for partitions where there is direct pig contact. Wire mesh divisions may be used for smaller pigs but a 20 mm × 20 mm or 50 mm × 10 mm mesh would be required for pigs weaned at 4 kg or less.

Gates are a point of direct pig contact, so sturdy construction is a prerequisite. As well as acting as a pen/house boundary they may also be used to effect indirect stimulus contact (i.e. boars:sows tactile divisioning in a service area). Where contact is desired tubular divisioning at 125 mm centres in a vertical form is recommended. Mesh divisions, even with support at 450 mm centres, is more likely to suffer from corrosion and pig contact. Sheeted gates demand heavier support, and whilst steel is a positive and warp-free cladding, it is prone to corrosion. Recycled plastic would overcome the corrosion risk but is heavy and does tend to warp.

Space boarding has been widely used in naturally ventilated structures but offers imprecise year round temperature control. The use of plastic netting on adjustable timber framework gives greater flexibility in this respect as the frames can be light in construction and easy to hinge or slide. The plastic netting must be fixed between a sandwich of battens and is substantially corrosion resistant, although some types may suffer damage from ultra-violet rays over a number of years.

Relocatable Structures

These are normally built upon heavy-duty steel skid units. Panel structures with their narrow spans can be of rigid construction. The benefits of such structures are that they are mobile and are thus suitable for an emergency, a tenant farmer or a site where planning approval for a permanently sited structure could not be expected. The cost per cubic area may be higher than for more conventionally constructed housing and the durability of relocatable structures in comparison with other building methods, particularly where larger stock come into contact with the fabric of such structures, is as yet unknown.

The Floor

Because of the direct and unavoidable contact between the pig and the floor the importance of a satisfactory surface is apparent. However this is a constant source of anxiety to producers. Production trends have led to an increased use of slatted floors and there is a wide range of materials available.

Solid floors have traditionally been constructed of concrete. There can be little doubt that this will remain the most common form of flooring for pigs. It is vital that close attention to detail is followed because of its effect on pig comfort and performance. These considerations should include the choice of components and care in mixing as well as in laying. Careful site preparation must not be overlooked as pockets of badly compacted sub-floor will lead to cracking and sinking.

To reduce costs many piggery floors have included small sized aggregate. In lying areas this should be discouraged as eventual wear exposes the aggregate, causes damage to the stock and encourages floor fouling. The preferred concrete surface comprises a sand:cement mix in the ratio of no more than 3:1, chosen with care to ensure that a suitable 'sharp' concreting sand is selected. Special additives are available to give the floor extra surface hardening and easy-clean properties.

The cement should be thoroughly mixed with no more water added than is necessary to give a rapid closing of the mix when lightly tapped with the sole of a boot. No extra water must be present but proper compaction must be allowed. In a piggery lying area the screeded surface should be as thin as possible so as to maximise the benefits of the insulation layer below.

However, screeds of less than 15 mm nominal thickness require much skill in laying and need special additives such as rubber latex. In most cases a 20 mm nominal thickness screed would be laid over the insulation layer with screeding boards set to provide suitable floor falls for surface drainage.

Joints should be clean and minimised as they tend to be a site of water penetration. The concrete itself should be lightly tamped, finished with a wood float, and a steel trowel used to give the desired degree of foot holding without abrasiveness. Floors laid in this way should have an adequate load-bearing capacity of 4.8 kN/m² (100 lb/ft²), as well as ensuring the required comfort and cleaning properties, *provided* that the floor is kept covered for a minimum of seven days to allow gradual curing and is not used for a further 20–25 days. In the United Kingdom the trade sponsored Cement and Concrete Association is a source of invaluable advice on floor construction and all products containing cement.

In the dunging areas where mechanical scraping or even hand scraping is practised, a different approach is required as greater load and wearing characteristics are required for the floor. A mix (UK specification No. C20P) comprising 9 parts coarse aggregate (20 mm maximum), 5 parts sand and 3 parts cement would be specified, laid to 100–150 mm thickness on a prepared base. Such an area in a piggery becomes gradually more slippery with the passing of cleaning/scrapers, and care must be taken to tamp such floors or to groove them in a way that will permit reasonable foot adhesion over years of use.

If care is taken in laying and choice of materials maintenance requirements will be low. However, because certain areas of pens (e.g. in front of troughs) become worn over time, careful cleaning, degreasing and patching may be necessary and can be successfully carried out by using epoxy resins, polyester resins, acrylic resins and polymer emulsions.

The slurry channel itself must be carefully constructed and finished so that no leaking of liquids into or from the sub-soil occurs. The former leads to blockage of the channel and the latter to excessive liquid output. The base of the slurry channel should be level in both planes and less than 30 m (100 ft) in length. If the building is longer a fall of 1:200 should be laid to the discharge point. Shallow slurry channels, those less than 1 m deep, do not always flush well if longer than 10 m (33 ft). In practice, although every effort to minimise water spillage from drinkers should be made to reduce the quantities of slurry which have to be handled, many slurry chambers clean more satisfactorily when extra liquids are added. Shallow channels may require supplementary water or re-circulated, separated, preferably aerated, liquid added to the channel to ease discharge and prevent build-up of solids. An alternative may be to operate a 100 mm high weir system over which excessive material will pass. This latter principle works less well on wider channels.

An alternative to sluice gate or suction removal of the slurry is to use below-slat scrapers. Access for maintenance and repair is not always easy to arrange and they do add considerably to capital costs. An open pipe system laid in the channel floor with 100–150 mm (4–6 in) diameter openings at 1 m (3 ft 3 in) centres is also used to arrange discharge from shallow channels.

Where sluice disposal is used the opening should be as wide as possible to encourage discharge of the solid fraction of the slurry.

The slatted area of a pen is similarly vital. In a farrowing pen, the aim must be to create a perfect balance between pig comfort (given the disparity in size between the sow and its newborn offspring), cleanliness and length of life. This has not yet been achieved with any type of material. Pigs in the weight range of 20 kg and over have a set of requirements much easier to satisfy. Measurements of pigs' feet and limbs have revealed that slatted floors should have gaps no greater than 10 mm yet, for adequate cleanliness, the void area should be close to 60 per cent. Materials likely to provide this combination of requirements are metal or plastic with close supports below. For pigs over 20 kg up to the adult phase there are less finite requirements in terms of minimum dimensions for foot and limb comfort.

Concrete slats have been subject to wide use and much abuse. In part the dissatisfaction has arisen from poor specification, poor materials and poor workmanship. However, like concrete floors, given good materials and operation, concrete slats may be as suitable as any alternative for pigs once 20 kg bodyweight has been reached. Ideally the slats will have 100–125 mm bearing surface with 17 mm gap (or 25 mm gap for adult pigs), with rounded edges for added comfort, easy cleaning and low likelihood of chipping. These slats are generally self-supporting up to 2 m in length and should have a long life, but are costly to transport and handle on site.

Metal floors have been used in many forms. Welded mesh was one of the earliest of floors used (75 mm × 10 mm × 5 or 6 gauge) and gives excellent cleanliness. The rounded welded bars make it comfortable, although sows in farrowing crates may find foot adhesion less than perfect. The welded mesh requires supporting at 200 mm for adult stock and almost double this spacing for pigs up to 20 kg. Its rate of wear is fairly high but capital cost is low.

Expanded metal has proven to be largely unsatisfactory in farrowing houses due to foot lesions and teat damage created by the edge of the mesh. It is, however, a satisfactory and cheaper floor for pigs in the 5–20 kg weight range.

Perforated or punched metal panels have been widely used due to their convenience because they are self-supporting and the ease with which a single, normally 200 mm wide, panel can be replaced. Damage to the feet of piglets can occur and the low void area inhibits free passage of dung. Some forms are slippery for sows in farrowing pens.

Woven wire flooring, normally using carbon or high-tensile steel has many of the benefits of welded mesh. It is more expensive but needs less support and appears to have double the length of life.

Plastic and polypropylene has a high piglet comfort factor but, as with many other forms mentioned, is possibly too slippery and insufficiently self-cleaning for use in a farrowing house, and these slats require close supports.

Mild steel, 'T' section profile or rounded bars, are capable of providing a non-slippery, highly durable flooring with dimensions suitable for all classes of pigs, and are self-supporting up to 1.5 m. They may initially be more expensive than some of the alternatives mentioned.

Cast iron slats are also made in panels with a 'T' section and rounded edges for animal comfort. Their merits and disadvantages are as mentioned for mild steel except that they may have an even longer life.

Insulating the Building

The value and necessity of insulation has long been accepted. However, good insulation without good ventilation control (see Chapter 3) is largely ineffective. Insulation will reduce heat loss through the structure, assisting in the maintenance of desired temperatures. The provision of better insulation allows higher ventilation rates to be used without lowering temperature levels and also increases the removal of moisture and gases. Insulation reduces the likelihood of condensation which would exaggerate heat loss and also increase the deterioration of the insulant itself, as well as the main structure of a building. Although structural heat loss is well understood, it must again be stressed that draughts must be prevented in order to stop a rapid increase in avoidable heat loss.

The minimum standard to apply to the roof and walls of a piggery, except kennel-type housing, is a U value of 0.6 W/m² °C. This can be met by the thickness of materials shown in table 4.1. It should also be noted that some of the materials require a vapour seal. Failure to protect insulation from the effects of moisture, from condensation, spillage or cleaning not only speeds up deterioration but reduces the insulation value. Most insulants are prone to damage by pigs so good protection is required. Damage by rodents has become a major problem in housed livestock farming and attention to sealing

Figure 4.1   The heat loss through a typical uninsulated structure

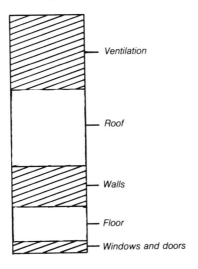

Ventilation

Roof

Walls

Floor

Windows and doors

Table 4.1   Summary of insulation materials to provide 0.5 W/m^2°C 'U' value

Material	State	Thickness required (mm)	Vapour barrier required	Cavity required	Lining required	Fire tolerance	Comments
Glass fibre	Mat	70	Yes	Yes	Yes	Class 1	Unpleasant to handle.
Expanded polystyrene	Board	60	Yes—if not foil faced	Yes	Possibly not if ceiling	Class 3	Very susceptible to rodent damage. Not easy to seal.
Extruded polystyrene	Board	50	No	Yes	Usually lined	Class 3	Easy to work with.
Urea formaldehyde	Foam	75	Yes	Yes	Yes	Class 3	Must be properly sealed to prevent leakage. Main use in walls.
Polyurethane	Foam	40	No	No	No	Class 1	Seals cracks and fissures, but has non-smooth finish.
Polyisocyanurate	Foam	40	No	No	No	Class 1	More expensive than polyurethane. Better fire tolerance.
Lightweight aggregate	Mineral mix	125	No	Yes	Yes	Class 1	Floor usage only

and to rodent and pest control should be given a high priority on any farm. Whilst table 4.1 gives a guide to the main properties of the materials available, the method of application and maintenance will also play a considerable part in the effectiveness of the material.

Special mention of floor insulation must be made. Lightweight aggregate with a monolithic floor screed laid in 125 mm nominal thickness should ensure that adequate 'U' value is achieved. The use of polyurethane panels in various forms can only be recommended where rodent protection is good. A flat sub-floor to prevent breakage of the insulation is necessary and a minimum of 35 mm screed is used to prevent impact damage. At this level of screed thickness the effectiveness of the insulation is reduced.

There is a gradual acceptance of the value of reflective linings both exposed and inside lined cavities to enhance insulation properties. Aluminium foil or brush-on reflective materials are most widely used.

The insulation of doors is often ignored and this should be specified in all conditions to prevent heat loss and condensation. Windows also require insulation by double glazing and it should be remembered that, although more expensive, the factory vacuum-sealed polycarbonate unit provides a better 'U' value and is less likely to deteriorate than a simple double-glass framework.

Airspace

Although not a consideration of structure as such, there has been an increasing trend, particularly in part or totally slatted structures, to reduce the cubic airspace within a building. The reasons for this, other than lower material costs, are that there is a reduced surface area through which heat can be lost and there is a smaller volume of air to keep warm, thus reducing problems of uniformity of house temperature. However, if the cubic air capacity is too low, slow airspeed at pig level is more difficult to achieve and this may lead to an unacceptably low rate of air exchange. This leads to high humidity levels and poor working conditions and is a common cause of poor pig performance if the cubic airspace to weight of pig ratio is less than $1 \text{ m}^3 : 25$ kg bodyweight.

Chapter 5

FEEDING SYSTEMS

THE CHOICE of feeding system will be influenced by a wide range of factors, any one of which may be given priority by an individual depending upon the prevailing circumstances. The form in which feed is delivered to the pen and its method of presentation to the pigs can have a direct impact upon the pen design, but the converse is also true as some layouts will preclude particular methods of food presentation.

SYSTEMS AVAILABLE

In general feed can be delivered to the pen in any one of three ways: manually, automatically or semi-automatically.

Manual Dispensing

As units have become larger, the movement and dispensing of feed by hand has come under closer scrutiny. The time taken and the sheer drudgery that can be associated with the task means that if hand feeding is to be the method used it must be given as much thought as a mechanical system.

Advantages
- It suits all shapes of buildings and layouts.
- It involves lower capital costs.
- It requires the operator to enter the house, so pigs are more likely to be inspected.
- It is easier to modify feed scales.
- It allows a variety of diets to be used in one building to suit various classes of pigs.
- Little maintenance is needed.

Disadvantages
- It is more time consuming.
- It increases the degree of physical effort required.
- Slower dispensing of feed increases animal agitation and excitement.

- It is not really suitable for wet feeding systems.
- It is less easy for management to be sure that the desired feed scales are being followed.
- Wider passages and doorways are needed to manoeuvre equipment.

To offset the disadvantages it is advisable to consider various points of unit and design detail. The aim with manual feeding systems must be to reduce the proportion of time taken *not* feeding, i.e. walking between feed store/bin and the pigs and gaining access to the feeding points. The siting of barn/bins relative to the pigs to be fed can speed up the task (see Chapter 2). Increasing the capacity of the feed barrow can reduce the number of journeys to the store. Size of doorways and passages can also make feeding more convenient. The use of step-in accessways, cut-away trough or hopper fronts can make dispensing and inspection of a trough or feeder quicker and easier.

Automatic Dispensing

Systems now exist for the complete automation of pig feeding from the raw material stage to delivery and dispensing of an exact amount to any given pen of pigs. Equipment is available to handle the feed in almost any form and to deliver it at pre-set intervals into almost any type of container for consumption by the stock.

Advantages
- It reduces the time taken and effort needed in feeding.
- Feed can be handled in any form.
- It may be programmed to incrementally adjust feed scales.
- It is possible to incorporate auto-feeding into a composite monitoring system via a computer.
- Quicker feeding reduces animal agitation.
- It is easier to check that predetermined feed scales are being followed.

Disadvantages
- Even the less sophisticated systems will increase capital outlay.
- All equipment has an inherent maintenance/depreciation cost.
- Mechanised feeding systems are dependent upon electricity supplies, so at times of breakdown routines can be disrupted.
- It requires more forward planning so that feed scales are adjusted when pigs are removed or where adult stock has reduced appetite.
- It is less easy to feed more than one diet in a single building.

The reliability of much equipment has now been increased so there is less reason for concern over maintenance. Larger units usually incorporate stand-by generation equipment to reduce the effects of mains failure. Where dispensers are filled immediately following the previous feed the operator has little chance to adjust feed scales for individual pens. However, it is a fact that wet feeding of pigs is gaining in popularity due to the cost advantages of feeding bulky materials and the sheer scale of many pig operations continue to make the advantages of automated feeding irresistible.

Semi-automatic Dispensing

By definition this is a combination of the dispensing methods previously discussed. In its common forms a semi-automatic system either delivers feed to the dispensing point leaving the operator to actually measure or deposit it into the pen, or does the opposite; that is to say the operator measures the feed manually into a container to allow him to dispense it quickly and conveniently at some later point.

Advantages
- It can be quicker than hand feeding and simultaneous, hand operated dispensing.
- It may require a lower cash outlay than fully automated systems.
- If hand delivery is used it allows finite adjustment of feed scales and diets when stock are at rest and there is more opportunity to accurately re-set feed scales.
- Although difficult to evaluate in cash terms, a simultaneous dispenser reduces stock excitement at feeding times.

Disadvantages
- Mechanical delivery of feed does not permit diets to be varied unless very sophisticated equipment is used.
- Mechanised delivery is also dependent upon electricity supply.

In certain situations, like a dry sow house, the benefit of being able to deliver feed, modify the feed scale, add wormers etc., and to 'silence' a large stall layout by simply operating a few levers has clear attractions. The converse adaptation of part automation, where feed is delivered to a pen by pump or auger and requires the stockman to operate a valve or switch does increase the opportunity to inspect all the pigs at feeding times.

MAKING THE CHOICE OF FEEDING SYSTEMS

The discussions above illustrate the circumstances which should be considered when making a choice of feeding system. Only guidelines can be given in a book because the final choice of system can only be made according to the circumstances of the individual unit, availability of various raw materials and the owner's and operator's personal preferences. The use of certain feed systems, such as ad-lib hopper feeding, for example, will influence the degree of automation used. The choice of hopper feeding can only safely be considered in certain clear-cut circumstances.

There can be no doubt that a degree of automation is necessary on the larger unit and that, if the advantages of wet feeding are to be capitalised upon, the benefits of mechanical mixing and delivery, if not dispensing, are overwhelming.

In addition to the degree of automation employed there are important considerations to make concerning the feeding system adopted. For example, it may be considered preferable to feed pigs ad-libitum, and the only appropriate system in these circumstances is to hopper feed. Hoppers lend themselves well to automatic filling by auger. Free-access feeding may be satisfactorily practised provided that good management ensures that no fouling of the hopper occurs and that the hoppers themselves are carefully designed to eliminate wastage and physical damage to the pigs.

However, ad-lib hopper feeding is not universally acceptable. Pigs from some genetic sources are not suited to free-access systems due to the levels of backfat which occur under such a system. Breeding stock will almost certainly consume more feed than is economic if ad-lib fed during pregnancy. Careful design and space allocation is also an important feature of hopper feeding systems because it has been noted that competition, siting and feed freshness can all combine to actually inhibit full appetite, and under adverse circumstances, hopper feeding will not lead to the most rapid growth. However given good hopper design, management and the appropriate genetic make-up, ad-libitum feeding has many advantages.

Controlled feeding requires either greater time to measure the appropriate feed amounts or greater sophistication, and therefore costs, to automate the system. Controlled or scale feeding also implies that feed is offered at intervals, and it is the combination of the control of quantity and the timing of feeding that gives rise to pig agitation.

Controlled feeding does allow economies to be made and a more satisfactory relationship between feed intake and efficiency, growth rate and carcass quality to be established. It is necessary with a controlled or rationed system to be ready to adjust scales according to the response of the individual pig or pen of pigs, and to vary the quantities given relative to changes in diet density. The choice of a feed scale must not be taken lightly, and must be interpreted with the appropriate flexibility on the part of the operator, ensuring that any pigs whose condition or growth is not satisfactory receive a suitably modified scale. For growing or finishing stock the most appropriate approach might be to use an age-related scale but to modify it according to the pigs' actual weight for age, so that the planned piggery throughput is achieved throughout to avoid overstocking of facilities.

More sophisticated systems of rationing and dispensing are becoming available. The 1980s have brought to the commercial stage the use of electronic identification for sows so that they may receive a given daily feed quantity from a feed station in the pig pen. The transponder allows a predetermined quantity of feed to be offered to individual sows depending upon their conditions and stage of pregnancy. Being linked to a computer the feed station and transponder can also be used as the basis for sow recording and monitoring. Feed dispensing according to the speed with which feed is consumed has been developed for finishing pigs to counter the variances between the speed of eating within a pen of pigs. By making separate, individual trough points available feed is dispensed in small quantities as previous amounts are eaten; the reduced competition which arises is claimed to lead to more even growth within a pen. These examples of added sophistication to a ration feeding approach obviously increase capital requirement but may have added advantages in terms of composite pig and feed recording and result in improved growth efficiency to help offset the extra costs.

Mechanisation also offers greater opportunity for increased frequency of feeding where controlled amounts are offered. Unfortunately few completely integrated trials or experiments have been conducted to guide the commercial producer on optimum feeding intervals using conventional trough or floor feeding methods. However, within reason more frequent feeding:

- Increases feed intake *provided* that intervals between feeds are regular. On-farm observation suggests that pigs fed at 3 × 8 hour intervals rather than 2 × 12 hour intervals may consume up to 5 per cent more feed and grow faster.
- Reduces variation in growth. This is because the opportunity for the more aggressive pig to dominate the timid eater is reduced if less is offered at any one time.

It must be stressed that greater frequency of feeding has never been shown to increase efficiency where wet feeding is practised. Trough capacity will mean that twice-a-day feeding is normally used and this ensures freshness.

Trough Feeding

Most breeding stock will be fed from a trough of some kind. In farrowing houses and service areas there is a distinct advantage in the use of a trough to hold the relatively large quantities of feed which should be offered at these stages. Trough feeding for growing/finishing pigs places some limitation upon pen shape due to the essential feed space allowance required. Clearly, if wet feeding is to be practised, a trough arrangement is essential. Trough feeding has several advantages:

- Wastage by fouling will normally be less.
- Fouling being reduced, trough feeding leads to larger intakes compared to floor feeding (by around 3 per cent) and better growth rates, particularly where meal is used.

Pigs have certain requirements for trough capacity and dimensions. Table 5.1 provides guidelines for the minimum allowances.

Materials
Those commonly used for troughs include the following:

Steel is widely used for breeding stock but should be of heavy gauge (12–14 gauge) due to relatively high rate of wear. It is not really suitable for fatteners.

Table 5.1 Minimum allowances of trough size and capacity

Class of pig	Trough width (mm)	Trough depth (mm)	Trough length per pig (mm)
Adult	300	150	400
Lactating Sow	450	200	450
Pigs up to 10 kg	100	100	75
Pigs 10–25 kg	150	125	150
Pigs 25–50 kg	225	150	200
Pigs 50–75 kg	225	150	250
Pigs over 75 kg	300	150	300

Salt-glazed is very hard wearing but heavy to position initially. It is essential to use strong mortar in joins to prevent leakage and weakening of the trough.

Concrete has a relatively high rate of wear which makes it unsuitable, although special coatings of polyester and acrylic compounds will increase longevity.

Glass reinforced concrete (GRC) has a much extended life over normal concrete.

Plastic is easy to cut and handle and, if carefully protected and mortar-bedded, will give long usage. It is widely used for small pigs.

Glass fibre is expensive to mould so is only normally used for weaners.

Polyester resin is very hard wearing but has higher capital cost.

Floor feeding

Floor feeding became widely adopted when part-slatted floor arrangements became more commonly used and the need for economies of capital increased. This method of feeding has several advantages:

- Its low cost.
- Higher stocking rates may be used.
- It does not dictate the use of a particular shape of pen.
- It helps to promote clean lying area habits, particularly where meal is fed.

Despite these advantages many users of floor feeding find it difficult to achieve feed intakes consistent with optimum growth rates. Losses of feed by trampling and spillage into the dunging area means that FCR is also often worse than for hopper or trough feeding. Floor feeding cannot be recommended where optimum growth and feed efficiency are the aim.

Wet feeding

As previously mentioned, the renewed awareness of the pig producer to the value of using a range of liquid and pulped raw materials is increasing interest in wet feeding. Wet feeding offers a flexibility in raw materials which no other system can provide. It also reduces the amount of dust and the associated losses which arise from this.

Where conventional cereal and protein feed is used the merits of wet feeding compared to dry feeding are still attractive.

Advantages
- The feeding of meal mixed with water improves feed conversion and growth rate.
- It is not necessary to mix raw materials prior to adding to the water (although it may be more convenient to do so).
- The system is normally cheap and simple to extend.

Disadvantages
- Young/smaller pigs are not well suited to wet feed under about 35 kg unless a thicker mix can be used, as the water volume normally required to move feed around a pipeline system can cause appetite to be satisfied before optimum feed intake has been achieved. A ratio of meal:water of 1:2.7 is common whereas, for small pigs a ratio of 1:2.4 would be preferable.
- Increased gut fill means that the killing-out percentage may be reduced.
- Gradings may be poorer; this is a reflection of faster growth and may not be economically significant if FCR is satisfactory.
- There is a risk of freezing if there is inadequate protection from frost.
- Blockage of pipeline may occur if:
    A grist size of more than 1/8 in (3 mm) is used.
    Pipelines are not flushed clear regularly.
    Mixes are too thick (i.e. ratio of meal:water is too low).
    There is a drop between the pipeline and feed valve allowing settling out.
    Pump capacity is inappropriate for the sizes of the pipeline and the length of distribution. As a guide a 50 mm diameter line on a ring circuit of up to 640 m should create no problem to an installation engineer.

A compelling advantage of pipeline feeding is its flexibility in the range of raw materials which can be used, even allowing for difficulties in storage associated with such products, e.g. the irregularity of supply and difficulty in accurately assessing their feed qualities.

Mixing Systems
There are three main types of mixing systems for wet feed:

Batch mixing where a predetermined quantity of feed is calculated and then mixed. This can be automated and the quantities of raw materials are either volumetrically measured or, in the most sophisticated form, are weighed using load cells. Batch mixers give a greater opportunity for varying diets for pigs at different stages.

Automated, repeat batch system uses the same principles as the batch system but the smaller capacity mixer means that the delays between each batch will extend feeding time. This method gives more opportunity to feed more than one ration.

Continuous mixing and distribution may be cheaper to install but offers less flexibility in the use of different feeds.

Feed is most commonly distributed using centrifugal pumps although scroll and stator types may also be used. Some systems use a compressed air as the distribution system. PVC piping is cheaper to install provided it is not suspended at high levels where supports at 450 mm–600 mm (18–24 in) centres will increase the cost. It is much less corrosion resistant than steel which is the main alternative. Steel pipes need supporting if used at high levels at 3 m (10 ft) centres.

The most common systems of frost protection are to lag pipes with an insulant or to simply drain the system. Low wattage electric cables are also used.

Some modern systems exist where feed is delivered in a dry form and is steeped in a container at the pen before subsequent dispensing in a wet form. This allows simultaneous feeding in a wet form and helps to overcome some of the problems of wet feeding such as the thickness of the mix used and conveying drawbacks.

Dry Feeding

It is also true to state that dry feeding of cereal-based diets has its merits. This is particularly so where suitable grains are grown on a farm or nearby and where production is geared to precise nutrient allocation rather than out-and-out feed cost per kg gain and exploitation of available by-products. Dry feed is truly flexible in that it is not necessary for a trough to be available as in wet feeding. Moreover mechanisation of feed distribution and dispensing is normally much less expensive than in wet feed. In addition it is difficult to arrange for simultaneous dispensing of a normal pipeline feeding system whereas dry feed dispensers can be mechanically or manually operated to silence very large numbers of pigs simultaneously.

Dry cereal-based diets in meal form are cheaper to produce but are less efficient compared to extruded products (pellets/cubes). Trials show conclusively that improved feed intakes, feed conversion ratio, growth rates and reduced dust levels can be attributed to pelleted feed. The heat generated in the pelleting process tends to enhance palatability and reduce the remote risk of bacterial contamination. Home mixers may be reluctant to install pelleting plants, however, due to their high capital costs and the unavailability of steam to form good pellet quality.

To attempt to obtain some balance between capital costs and the performance merit associated with wet feeding, some operate a compromise system of delivery and dispensing feed in a dry form and allowing pigs to wet it to their own requirement by siting drinkers low over a trough layout.

Delivery Systems

There are a number of delivery systems available for distribution of feed in its dry form. These are:

Enclosed disc and chain/wire. This system is fast and reasonably silent. It tends to break down pellets to some extent and is unsuitable for larger (10 mm diameter) pellets.

Continuous spiral auger. These augers have a long delivery capability but continuous rotational action increases metal-to-metal wear although PVC may reduce the rate of wear.

Conventional worm auger. Although cheaper they do not work satisfactorily at less than 45° as they break down feed and have a high rate of wear.

Chain and flight. This system is quiet, gentle with feed and virtually resistant to blockage. It is less flexible than the other systems however and has a slower delivery rate as well as being rather more expensive.

Open reciprocal. This has all the disadvantages/advantages of chain and flight and, in addition, is quieter and almost maintenance free.

Conclusion

This chapter highlights the fact that there is no one correct choice of feeding system and a final choice will depend on the particular circumstances of the farm.

Because the feeding system determines the pen shape and this, in turn, can influence house layout it is essential that consideration be given to the long-term raw material availability prior to committing oneself. One particular layout might not be easily adapted for a change of feeding method. House layout is stressed in later chapters particularly those dealing with grower and finishing pig housing.

Chapter 6

WATER

THE AVAILABILITY of adequate water supply is too frequently assumed to be satisfactory simply because an automatic drinking valve is installed in a pen. To make such an assumption may be a most serious mistake because water intake may be limiting the full performance potential due to a variety of factors: unsuitable type of drinker for the class of stock housed; inappropriate siting of the drinker; insufficient drinker points for the number of pigs penned; inadequate water supply or rate of flow.

The provision of the correct supply of water demands a degree of consideration equal to that of any other housing component. Where water supply is inadequate, appetite may be restricted, giving rise to poor performance. At worst, cases of salt poisoning may be observed in the pigs, or badly installed or unsuitable drinkers may give rise to excessive spillage, creating problems with environment control and increasing the cost of effluent disposal.

Needs of Pigs

Ascertaining the precise needs of the various classes of pigs has given rise to a certain amount of disagreement amongst researchers. A general guideline shows that on a weight basis, pigs require water in a ratio of 3:1 for each part of dry feed. However, the type of housing, temperature, humidity and the constituents of the diet may all influence the demand. It has been noted that higher amino acid inclusions in diets may increase the daily water requirement. This may be due to the regulatory influence of the kidneys. There can be a great variation between individuals and this is particularly noticeable amongst adult animals. Table 6.1 provides an approximate guideline, but must be interpreted with care as it assumes that no restriction of supply occurs at the time of peak demand (which is soon after feeding) and that there is no other major impediment to the pigs' drinking preferences.

Safeguarding the Rate of Flow

When installing a water system it is vital to safeguard the rate of flow. Although this is a somewhat arbitrary figure, observations of pig behaviour

Table 6.1   Daily water requirement for various classes of pigs

Class of pig	Daily water requirement (litres per day)
Maiden gilts	5–6
In-pig sow or gilt	5–8
Lactating sow or gilt	15–30
Boar	5–8
Pigs up to 10 kg	1.2–1.5
Pigs from 10 to 25 kg	2.25–2.5
Pigs from 25 to 50 kg	3–5
Pigs from 50 to 100 kg	6–8

show that the animal may spend up to twenty minutes drinking during a 24-hour period. So first it must be ensured that the rate of flow or 'head' is adequate to supply such demands. This can be measured in existing systems by timing the rate of flow in a measuring container. In a new building it is appropriate to consider the following factors:

- The pressure is sufficient from the mains supply tanks. This is influenced by a combination of tank size, its elevation above the height of the drinkers and the length of the distribution pipeline. Table 6.2 gives a guideline to the pressure from supply tanks positioned at various heights. It should be noted that various statutory limitations may be imposed by water authorities regarding the use of water. Few allow animal drinkers to be attached to a direct mains supply and so a header or break pressure tank is normally needed. It is then the height of this tank which determines the pressure in the water feed line.

Table 6.2   Pressure at differing heights of supply tank

Height of base of supply tank above drinkers (Metres)	(Feet)	Water pressure (lb/in^2)	
1.5	5	2.16	
3	10	4.33	Low pressure
6	20	8.66	
9	30	12.99	
12	40	17.32	
15	50	21.65	High pressure
18	60	25.98	
21	70	30.31	

• The capacity of the supply tank is an important consideration in providing security against mains failure. An on-site standby reservoir is of great benefit because if supply is interrupted due to unforeseen circumstances it can provide some water until normal supply is restored. Ideally, this reservoir would have the storage capacity to supply a unit for seventy-two hours. In some circumstances pressure may be too high in which case a pressure-regulating valve should be positioned in the house supply lines to protect the drinkers and fittings.

• If the circumstances of the farm or house prevent the supply tanks being sited to provide an appropriate rate of flow, then the installation of circulatory pumps in the water system between the supply tanks and the drinkers should be considered. Normally these are electrically driven and are activated by water demand. As pigs drink and water is drawn along the pipeline, the pump automatically cuts in, ensuring constant pressure throughout the house/farm drinking system. This is particularly valuable in large houses which are fed at intervals and have cyclical peak demands for water during the day. Alternatively, constantly revolving pumps on a ring circuit from and to a main tank supply may be used.

• Size of supply line may also have some effect upon rate of flow. If the internal diameter is too small it can inhibit rate of flow, particularly at some distance from the supply tank. As a general guideline the main feed line should be rather larger in diameter than the 'droppers' to the drinking points themselves to maintain even supply throughout the system. In certain installations it is necessary to clean the line occasionally as part blockage may arise and reduce rate of flow. Blockage can be caused by sedimentation in the line or scaling caused by limestone deposits.

• Most types of drinkers can be supplied with differing valve fittings to suit available water pressure. Where supply is relatively low, less than 15 lb/in^2, a low-pressure valve fitting would be specified.

Siting the Drinkers

This can influence the efficiency of both the water system and the house itself. The drinking points should generally be positioned in one of two positions:

—in the dunging area
—over the trough.

There are some important points to follow regardless of the site selected:

• The position of the drinkers should not be a limiting factor in water uptake.
• Any spillage should not produce excessive wetness of floors, particularly in lying areas. This would cause reduced level of comfort and accelerate heat loss from a building.
• The drinking point should not be such that it is easily fouled by the pigs or damaged by the operation of gates or passage of vehicles.

Where the houses have troughs as a part of the feeding sytem, it may be beneficial to position drinkers low over the trough, enabling pigs to operate the drinker as they feed and so create less disturbance at feed time. Any spillage would be held in the trough which should reduce floor wetness, water wastage and act as a reservoir of supply for any individual pig reluctant to operate the drinker itself.

Where no troughs are installed, or where there is concern about increased risks of feed becoming stale and unpalatable in its dampened form, the drinker should be positioned in the area of the pen designated as the dunging zone. In most house layouts the dunging area is often the furthest point from the operator's inspection passage so it is important to ensure that the water system is sited to give satisfactory operation.

As pigs tend to dung more in the corner of a pen it is advisable to position drinkers away from that point and reduce fouling and obstruction of the drinker. If pen dimensions permit, the drinker should be at least 1 m from the pen corner with a gap of at least the length of the pigs to be housed between drinkers.

Choice of Drinker

A common fault in watering systems is to assume that one type of drinker is suitable for all circumstances. This is certainly not the case. The type of drinker should be chosen with care taking into consideration the pig category, housing system and pen shape. Although one drinker per pen is common, two drinker points safeguard breakage and reduce competition and aggression.

Bowls with Float-valve
These have the merit of providing a constant reservoir and so suit most water pressure supplies and pig categories. However, they are easily fouled by food from the pig's head and by bedding and excreta. The latter two problems may be solved by raising the bowl above floor level on a plinth with a step to allow the pigs to reach it. However, the problem of food spillage remains and will require that the bowls are cleaned daily by the operator. So these are unsuitable for siting near to a trough in a lying area such as a farrowing crate. Due to their ready water availability, one bowl of this type would suit up to twelve pigs in a pen.

Flap-operated Bowls
These use the flap mechanism to prevent fouling by bedding and excreta, while allowing flow of water into the bowl. The shallow reservoir of water also leads to less feed building up in the bowl than in the float valve type, so less spillage would be expected with this type of bowl. Ideally, they are sited on a kerb or raised to give a 150 mm (6 in) height of the bowl lip above the floor level for pigs over 10 kg and 75 mm (3 in) for pigs below this weight. This bowl-type is suitable for up to ten pigs in a pen.

Nipple-type
These normally have a ball valve which is removed from its seating when a pig operates a central nipple-type plunger, allowing water to flow. Provided that they are fitted according to the manufacturer's guidelines and the supply is kept free of foreign material, they are quite leak-proof and easily operated by almost all classes of stock. They have the added merit of low unit cost and easy installation, working best from a 45° wall bracket. However, they are prone to spillage when in use and are easily knocked by pigs moving past them. Because water flows easily from them they are frequently 'played with' and may be a cause of excess pen wetting or slurry output. This type of drinker should be fitted to make the pigs which are to use them stretch slightly upwards. The tip of the drinker should be positioned 200 mm (8 in) above the floor for pigs up to 10 kg and up to 600 mm (2 ft) above the floor for pigs at 100 kg. In its vertical position (the nipple point downwards) it is suitable for use above a trough with the tip 150 mm (6 in) above the base of trough. A maximum of eight pigs per nipple drinker is considered safe.

Bite-type
These have various flow control mechanisms but rely upon the pig enclosing the valve in its mouth to reduce accidental spillage. Some are best installed on a 90° bracket, others work equally well from a 45° fixing. These valves are not universally favoured by all pigs. Some simply fail to operate them satisfactorily, possibly due to some physical defect or mouth damage. In addition their unit cost tends to be higher than nipple or spray types, although this remains a small percentage of the total housing cost. Fitting heights and pigs per drinker points are similar to the nipple-types.

Spray Drinkers
These are designed for use above a trough where the pig operates a valve with its snout and drinks from the trough rather than direct from the drinker as with the bite-type. These are normally very leak resistant if correctly fitted and foreign matter is kept from the supply line. These should be sited with the tip of the drinker 150 mm (6 in) above the base of the trough. For sows with individual troughs, the drinker should be 100–150 mm (4–6 in) in from the side of the trough for easy operation without discomfort and continual operation when the sow is eating from the trough. In finishing situations one spray drinker per two pigs is recommended.

Table 6.3 provides some guidelines to appropriate matching of drinkers to category of pig or housing system.

Maintenance of the System

Reference has been made previously in this chapter regarding the need to check regularly that adequate flow is maintained. It is always advisable to check this at the drinker points furthest from the point of supply fifteen minutes after a feed when demand for water is normally at its peak. A simple check is to use the daily requirement figures indicated in Table 6.1 and to

Table 6.3   Suggestions for choice of drinkers

Class of pig	Housing system	Suggested drinker type
Adult	In groups	Flap-operated bowl or bite-type
Adult	Individual	Spray or nipple positioned vertically
Weaners	Slatted	Flap-operated bowl or nipple
Weaners	Bedded	Flap-operated bowl
Grower/finisher	Slatted without trough	Bite-type
Grower/finisher	Bedded without trough	Flap-operated bowl or bite-type
Grower/finisher	With trough	Spray type over trough

multiply by the number of pigs in the pen. It might be expected that pigs will only be prepared to compete for water for around twenty minutes during a one-day period so the necessary rate of flow can be calculated. For example:

- In a pen used for pigs up to 50 kg in weight each pig may need 5 litres/day.
- Assume that this water must be consumed within twenty minutes the minimum rate of flow gathered from the drinker point should be 5 litres ÷ 20 = 250 ml drunk in one minute.

The supply tanks to the system should be covered to prevent contamination and the subsequent growth of algae which is not only a potential medium for water contamination but may also give rise to blockage of the system. The inclusion of proprietary sanitising material will increase the efficiency of the watering system. Dispensers exist to regulate the quantity used to keep the circuit free from contamination.

Many drinkers, particularly flap-operated bowls and bite drinkers, have gauze filters fitted and these should be occasionally checked to ensure that they are free from blockage. Checks on the drinker valve itself are particularly recommended when a new circuit or drinker is fitted. Maintenance frequently admits foreign matter into the supply pipe and this tends to settle out at the lowest point—which is a drinker valve.

Water Medicators   When it is necessary to provide medicines or additives to a proportion of a herd in-line medication is attractive. However, many drugs are not suited for this use and many medicators do not safely handle all additives. The inclusion of water-based additives demands that certain principles are closely adhered to.

- Extreme care with hygiene must be maintained and the use of sanitisers to ensure that the watering equipment and the line are bacteria-free is vital.
- The material used must be water soluble.
- Vane-type pumps may be more quickly corroded than peristaltic types by certain additives.
- The chances of blockage of supply lines and drinkers is increased where additives are used and extra attention to regular maintenance is required.

Finally, it must be stressed that careful thought should be given to all aspects of water supply and that individual farm or house circumstances may make certain methods of watering unsuitable. A drinking system will give better results if it is regularly inspected and properly maintained.

Notes on the Use of Chapters 7–12 on Housing Design

THIS SECTION of the book establishes the critical requirements of the various classes of pigs and offers guidelines upon which a choice of system or layout might be made or compared.

Clearly, given the strategic considerations implicit in choice of system discussed in Chapters 2–5, certain layouts described would be immediately inappropriate in certain farm circumstances and this would narrow the choice.

In addition to the critical elements of design a number of pen/house layouts are shown with the relative merits and demerits set down to allow comprehensive consideration to be given in relation to the requirements of the pig and convenient, caring operation. Wherever possible the critical elements of design for each layout are indicated to avoid perpetration of the common errors associated with 'bad copies' of a basically sound design.

Many of the different layouts have examples of typical buildings shown in the illustrations; whilst some differences occur compared to the descriptive text, they give an indication of the type of design referred to.

It should be noted that some of the dimensions are 'rounded' and that space allowances vary for the type of feeding system used. The dimensions given are guidelines only and are not suggested as absolute standards but as illustrations of a particular system or layout.

Because inter-related factors such as siting of services etc., can make a large impact upon labour usage, clearcut staffing needs cannot accurately be given. However, in general, bedded systems may increase daily routine but slurry removal has its 'weekly' or occasional peaks. The carting and stacking of straw also imposes a seasonal peak of labour demand.

Chapter 7

HOUSING THE GILT AND DRY SOW

APART FROM the thorny debate concerning the degree of restraint applied to pigs and the impact of this on the choice of system it is necessary to separate this section of housing the non-lactating breeding female into three broad areas: the gilt; service time; pregnancy.

HOUSING THE GILT

It is normal to consider gilt housing from approximately one month prior to first service from a weight of 90 kg onwards.

Critical Considerations

- The number and dimensions of pens should be calculated according to the scheduling recommendations outlined in Chapter 2.

- Monitoring suggests that the housing of gilts in small groups (up to twelve per pen) is advisable for the management of growth and stimulus of oestrus.

- Where critical temperatures are not maintained within the bounds outlined in Chapter 3, environmental provisions will tend to depress reproductive efficiency. In particular if unbedded penning is used, attention to insulation and ventilation standards become vital.

- The spatial requirements of pigs have been subjected to close scrutiny by scientists and it is possible to produce theoretical foundations upon which design considerations can be made. These calculations take into account the size of the animals concerned, their contact with other pigs (i.e. group size), feeding and watering provisions and temperature control. The farmer needs to have some yardsticks in mind when considering pen size, and for gilts these are:

Lying area of 0.5–1.0 m² (5.4–10.75 sq ft) up to first farrowing, increasing by 20 per cent if UCT cannot be maintained.

Dunging area if slatted, approximately 25 per cent of lying area shown; if bedded, normally necessary to be in a 1:1 ratio with lying area, depending upon the muck removal method.

- Due to the variation which will exist between individual pigs, a degree of individual feeding should be considered to reduce the amount of the discrepancy in body condition which might otherwise arise and which can cause subsequent management problems. Floor feeding cannot be recommended; if a communal trough is to be used, it is advisable to incorporate short dividers to protect the less aggressive feeder. Individual feeders should be no less than 1.8 m (6 ft) in length overall, with a trough of sufficient capacity to allow spill-proof, once-daily feeding. A solid splash-plate on the gate front reduces the risk of gilts spilling feed. Side opening gates make removal of gilts without feed spillage easier. The rear gate should be solid to prevent damage to gilts' vulvas when being opened. It should be possible to open this rear gate mechanism from the front feed passage or the rear dunging area for convenience. The preferred gate action has a centre pivot so that the gate lifts up and over the feeder itself.

- It is necessary to consider the housing type when deciding upon the width of pens and passages. Some basic considerations are:

   Pen widths. Minimum wall length dimensions of any area should be no less than 1.8 m (6 ft) to allow injury-free movement of stock in a free-access area.
   Pen doorways. A minimum of 900 mm (3 ft) is recommended in group-housing layouts.
   Passageways. For feeding a width of 900 mm (3 ft) may be adequate, but where pig movement is also involved 1.05 m (3 ft 6 in) is recommended and any manoeuvring of stock is made easier if the passageways are 1.2 m (4 ft) wide.

A Critical Appraisal of Some Common Gilt Housing Layouts

Due to the obvious need to stimulate oestrus it is unusual in practical circumstances to consider gilt penning separately from sow penning. However, due to the need to reduce duplication in the book, gilt penning on its own is considered in this part of the chapter. In the next part of the chapter considerable stress is placed on boar proximity and ease of pig movement. These important considerations should be borne in mind when these first layouts are studied.

As in other sections of these Layout Design chapters some discrepancy in the floor area per pig may be noted between the various layouts. In some part these are due to inevitable 'rounding' of pen lengths and widths and, in other cases, to the difference in feeding or dunging methods used. Occasionally, the necessary minimum passage width to allow the use of a conventional tractor initiates the need for more generous dimensions or for a particular pen shape.

In certain cases typical materials are referred to although suitable substitutes exist. For example, in Layout 1, space boarding is indicated above the kennels but other materials such as nylon mesh on a suitable frame (see Chapter 4), could be substituted without affecting the operational sense or effectiveness intended. In the same way the depth of the slurry channel referred to in Layout 3 is not critical and a flushed system or the use of a weir would allow much shallower channels to be used.

Thus, these gilt-penning layouts and those which follow in subsequent sections are intended to provide guidelines to the kinds of layouts that might satisfactorily meet the critical considerations summarised at the beginning of this chapter. The layouts shown would meet the essential requirements but do *not* exclude the use of other designs. Again, dimensionally they would be subject to some modification but only if the critical requirements for stock and stockmen are met.

Few references to the siting of drinkers or water supply are made in Chapters 7–12 and attention is drawn to Chapter 6 where specific recommendations are made as to the choice and positioning of drinkers. It is, perhaps, particularly important to draw attention to the special needs for frost protection of water supplies particularly within those buildings where only parts may be insulated (such as the typical kennel layout illustrated in Layouts 1 and 2). Typically, supply to the drinker would be made via the lying area to reduce the risk of frost damage.

Finally, the number of pigs shown per pen is not critical except for the observations already made in Chapter 3 concerning group sizes and the impact of too many pigs per pen upon performance.

LAYOUT 1

Kennel and Tractor-scraped Yard with Individual Feeders (Six Gilts per Pen)

Important Considerations

The lying area must suit the needs of the six gilts. Layout 1 gives approximately 0.8 m² (8.5 sq ft) per gilt.

The dunging area must be wide enough for a standard tractor. The minimum recommended is 2 m (6 ft 6 in), see Chapter 4.

Tractor-cleaned dunging passages should be set down between kerbs to ease cleaning without damage to gates and the fabric of the penning and to create a natural drainage point in the pen. These kerbs should be 100–150 mm (4–6 in) in both vertical and horizontal planes.

The feed passage should be 0.9–1.2 m (3–4 ft) to make feeding and pig movement convenient.

Operating the System

The gilts are placed in their pen either by closing other animals into the kennels and driving gilts along the dunging passage or via the front-opening individual feeders from the access passage.

The gilts are enclosed in the kennel on alternate days to allow tractor cleaning of the passage. Damp bedding is thrown from the kennel by hand and fresh straw added to the dunging passage from the kennel top store or tractor/hand barrow.

Feeding is once a day by hand.

The gilts are removed, preferably through the front-opening feeder, into the access passage to be taken to the boar for observation/service.

Merits of the System

There is excellent control of feeding due to the use of individual feeders.

Muck removal is convenient.

There is no need for a specialist building, although good temperature control must be provided within the kennel.

Disadvantages

A large area is required per pig. Layout 1 shows that, including a 'share' of the access passage, a floor area of 3.03 m² (32.6 sq ft) per gilt is needed.

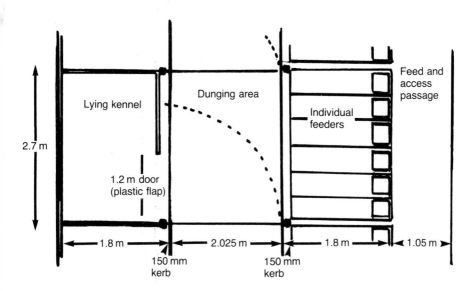

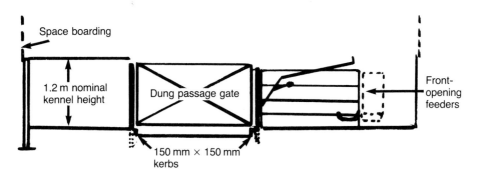

LAYOUT 1   Gilt penning – kennel and tractor-scraped yard with individual feeders (6 gilts per pen)

Unless boars are in adjacent pens there may be no nose-to-nose contact between the gilts and boars.

Actual inspection of the gilts is difficult from the access passage except at feeding time.

LAYOUT 2

Voluntary Stall or Cubicle and Tractor-scraped Yard (Four Gilts per Pen)

Important Considerations

The lying/feeding stall or cubicle must be of appropriate length and width.

To prevent bedding being scraped out of the lying area a profiled kerb should be used to help retain the straw.

There should be a slope from the trough of approximately 50 mm (2 in) over 900 mm (3 ft).

The trough should either be depressed to allow the gilt to utilise the full length of the standing or raised by 150 mm (6 in) to allow the head of the reclined gilt to be placed beneath the trough.

If an uninsulated structure is used it is essential that the lying/feeding area be turned into a cubicle by covering over that section of the pen and incorporating a flap for feeding/inspection of 600 mm (2 ft) depth. A cubicle *must* be draught-proofed at the front. Little opportunity for a removable front in such a layout exists, although it is possible in an open, insulated pen design. It is advisable to make provision for the suspension of plastic flaps at the rear of the cubicle to help lying area temperature control.

Operating the System

The gilts are fed daily from the access passage and can be enclosed by shutting the dunging passage gate across the rear of the feeding stalls.

The gilts are enclosed into the lying stall/cubicle by the dunging passage gate to allow cleaning on alternate days, following which, bedding is added.

In a cubicle system a second passage is required parallel to the dunging passage to allow access via a handgate to remove gilts for observation/service. In the voluntary stall within an insulated building, the second passage may not be needed as access to the gilts could be gained through a front gate on the feeding/lying stall.

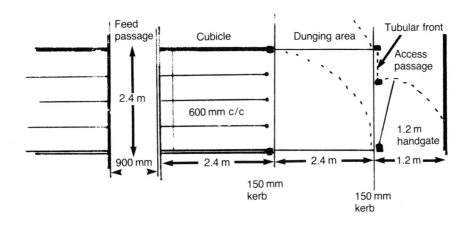

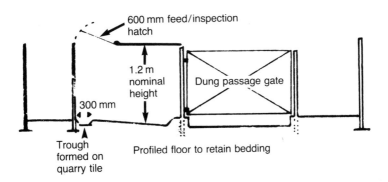

LAYOUT 2   Gilt penning – voluntary stall or cubicle and tractor-scraped yard
            (4 gilts per pen)

Merits of the System

The non-cubicle form permits excellent observation of gilts from the access passage.

It requires similar space to Layout 1.

The cubicle form does not require a specialist structure.

It allows excellent control over feeding of the individual gilt.

Disadvantages of the System

Access to the gilts by the stockman is not easy unless a second passage alongside the dunging area is provided.

Observation in any cubicle layout is always difficult.

Occasionally gilts will attempt to use the standings communally rather than individually.

Boar contact problems are the same as for Layout 1.

LAYOUT 3

Part-slatted and Trough-fed Layout within Insulated Specialist Building (Six Gilts per Pen)

Important Considerations

The lying area must comply with the space requirements indicated.

The slatted portion must be no less than 900 mm (3 ft) wide.

The trough should be 450 mm (1 ft 6 in) in length with dividers protruding 300 mm (1 ft) from the leading edge of the trough. The trough itself should be 300 mm (1 ft) in diameter.

It is vital in unbedded layouts to provide good temperature control.

Operating the System

The gilts are placed in the pen and withdrawn for observation or service via the handgate into the access passage.

Feed is deposited manually or automatically into the trough which should have dividers to within 50 mm (1 in) of the trough base.

Slurry is removed in the appropriate manner from outside the house.

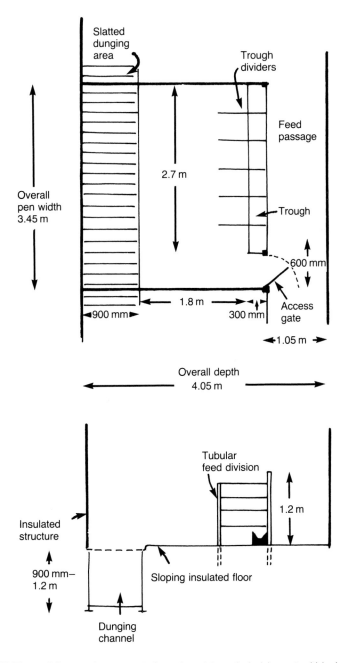

LAYOUT 3   Gilt penning – part-slatted and trough-fed layout within insulated specialist building (6 gilts per pen)

Merits of the System

The muck removal system reduces daily routine requirements.

Excellent stock observation is possible.

A reduced floor area is required: 2.03 m² (22 sq ft) per gilt, including sharing the access passage.

Disadvantages

There is reduced control over the sharing of feed and prevention of bullying.

SERVICE ACCOMMODATION

This section deals with the juxtaposition of the boar and newly-weaned sows or gilts, not specifically with pen dimensions which are detailed elsewhere.

Critical Considerations

Pens should be laid out to afford the closest possible contact between the boar and the sows or gilts. All oral, visual, nasal and physical contact encourages rapid reproductive activity.

Easy sow movement is vital to facilitate checking routines.

Good stock observation also assists efficiency.

Good illumination is especially critical within a service area.

Although group housing of weaned sows may help to speed the onset of oestrus, care must be taken to protect the more timid sow from competition at feeding time.

The pen in which mating is to take place has several vital requirements which are discussed in Chapter 8.

Good temperature control is essential in attaining full performance. Ventilation systems should not excessively reduce the influence of boar odour on oestrus stimulation. Recirculation ventilation systems may be useful.

Divisions between boars and sows should ideally be of vertical tubing at 125 mm (5 in) between centres. Between adjacent boar pens solid walling must be at least 1.5 m (5 ft) above floor level.

It is preferable to provide sufficient penning for sows for approximately five weeks after weaning and gilt places for one month prior to and after intended service to allow convenient checking for any returns to service.

Due to the need to move frequently at this stage tie stalls or tethers are not recommended.

The ease with which stock can be inspected and moved for checking is critical to the success of a service area. Thus, secure yet easily operated gate fasteners should be given careful consideration as should the way in which gates may be opened to ease sow/gilt movement. Included in such considerations are the need to reduce any obstructions or protuberances (e.g. drinkers) which might increase the risk of damage and make serving of sows less convenient.

A Critical Appraisal of Some Service Layouts

Layouts 4–7.

LAYOUT 4

Group-housed Sows with Voluntary Stalls and Bedded, Tractor-scraped Passage (Five Sows per Pen)

Important Considerations

The sow lying/feeding stall should be a minimum of 2.1 m (7 ft) in overall length with either a depressed or a raised trough. Minimum stall width is 600 m (2 ft) with a 190 mm (7½ in) nominal clearance between the bottom dividing rail of the stall division and floor.

A profiled floor and the need for kerbs are discussed in the Gilt Section (see Layout 1).

A well-insulated building with accurate temperature control is needed.

Operating the System

The sows are placed in groups via the dunging passage.

The sows are withdrawn via the dunging passage into the boar pens which are in a row parallel to the sow pens.

Individual feeding can be automated. Boars are fed by hand in individual troughs.

Boars serve in their own pen which is cleaned by hand, throwing muck into the sow's dunging passage.

The vertical tubing dunging passage gate has a sleeved arrangement to allow the gate to be extended to enclose sows in voluntary stalls with locking, slide assembly.

Merits of the Layout

Excellent sow:boar contact is possible through the vertical tubular pen divisions.

There is an economic use of space and the opportunity to use a double row of sows layout (with boars fed from the dunging passage); or for served sows to be moved to a different layout on the opposite side of the access passage adjacent to the boars.

It allows excellent sow observation.

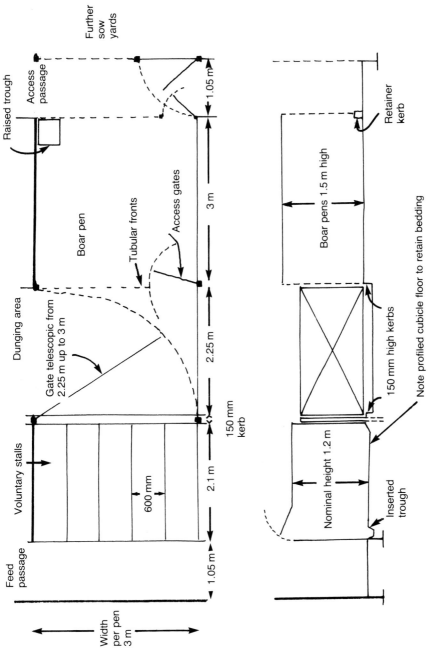

LAYOUT 4 Service penning – group-housed sows with voluntary stalls and bedded, tractor-scraped passage (5 sows per pen)

Disadvantages

Access to the sows is by enclosing adjacent sows into their voluntary stalls and this is relatively inconvenient where there is more than one group of sows to be 'tried' with the boars.

Feeding/access to boar pens is inconvenient using a single access passage layout.

Any cubicle/voluntary stall layout is not well suited to gilts.

LAYOUT 5

Alternate Sow and Boar Pens with Tractor-scraped Passages (Four Sows or Gilts per Pen)

Important Considerations

The dimensions previously given for tractor-scraped systems apply (Layouts 1 and 2).

This layout can be installed under a barn-type structure by enclosing the sow feeding/lying area into cubicles and covering the boar's lying area to a height of 2.1 m (7 ft).

Sow and boar pens are the same width to suit the dunging passage arrangement.

Operating the System

Sows gain access from a handgate into the access passage, which could service a second row of pens on this layout. This passage is also used for sow movement to boar pens.

Boars must serve in their lying area as tractor-scraped passages become notoriously slippery and unsuited to mating.

The boar pen has a trough for easy feeding.

Merits of the System

Sows are easily moved.

It gives excellent sow:boar contact.

Reasonably good observation is possible even if the sows' lying area is enclosed via the handgate from the inspection passage.

Disadvantages

Relatively large area is required due to the size of the dunging passage.

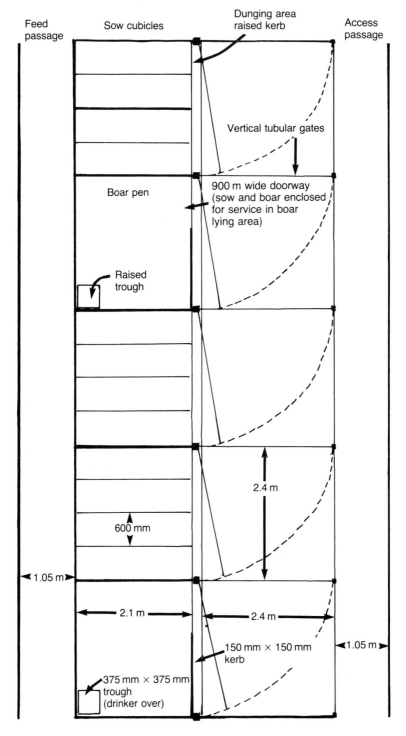

LAYOUT 5   Service penning – alternate sow and boar pens with tractor-scraped passage (4 sows or gilts per pen)

Any compromise on pen dimensions means that the boar's lying area, which is used for service, tends to be small. It may be desirable to provide a separate service area to which sow and boar are moved.

LAYOUT 6

Yarded Sows, Individually Fed, with Separate Row of Boar Pens (Six Sows or Gilts per Pen)

Important Considerations

Dimensions and critical considerations as detailed elsewhere in this chapter and Chapter 8.

Operating the System

The sows are placed into the pen either by enclosing other sows into their lying area, or via front-opening feeders which are essential in this service layout.

Sows and boars are fed from the same single access passage.

Mating takes place in the service pen between each pair of boar pens.

Merits of the System

A simple barn-type structure can be used due to the provision of a kennel for the sows and the boars.

A single access passage services all functions.

If the sows are retained in the feeder after feeding, movement to boars is relatively convenient.

A really good mating area can be provided.

Disadvantages

There is no physical sow:boar contact, although other sensory contact is good.

Sow observation between feeding is not easy.

A relatively large area is required per sow.

The boar pens require manual cleaning.

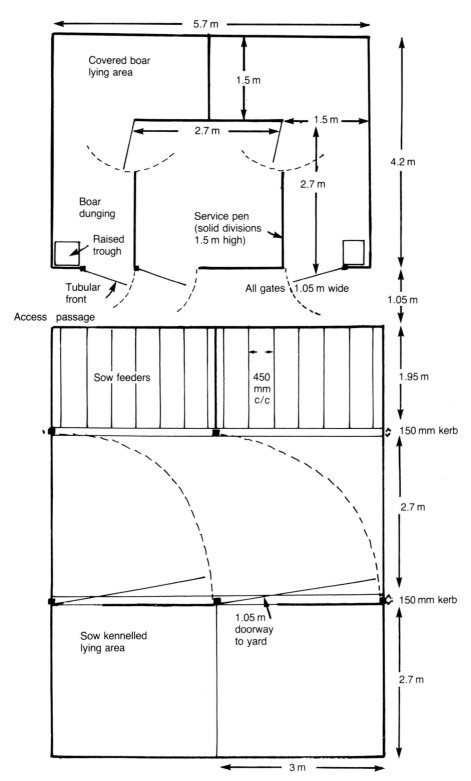

LAYOUT 6 Service penning – yarded sows, individually fed with separate row of boar pens (6 sows or gilts per pen)

Layout 7

Individually Penned Sows on Part-slatted System Opposite Boars

Important Considerations

The need to create adequate pen dimensions and suitable access ways is detailed elsewhere.

It is necessary to use front-opening stalls to permit easy sow removal for observation or service. Although top-hinged gates are convenient to fit, the need to feed newly-weaned sows generously and their short-term period of inappettance when in oestrus means that side-hinged gates are preferable to avoid feed spillage if troughs are positioned on the gates. Provision of water is easier when troughs are fitted *in situ*. If troughs are gate mounted, water points fixed to the gate front and serviced from a flexible hose are necessary. In such circumstances it is important to fit a drinker which forces the sow to operate it and then drink from the trough to avoid wetting of the floor. Ideally, the front-opening gates open from each side to allow sows to be conveniently moved in either direction.

Boar pen fronts should be of vertical tubing with solid walls between boars.

Good thermal insulation and temperature control is vital as this is in an unbedded system.

Operating the System

The sows are placed in the stalls via the narrower rear passage with movement boards used to allow them to be driven in the desired direction by one person.

The sows are removed from the front of their standing for observation/ service, and returned via the passages at each end of the building, or every 18 m (60 ft) down the building, running at right angles to the main feeding/access passages.

Boars serve in their own pens which require a degree of bedding—shavings, sawdust, or a similar material.

Boar pens are hand cleaned or scraped to a gulley over a slurry channel, although bedding can create difficulty in such circumstances.

Feeding has to be manual due to the front-opening stall.

Merits of the System

Only a small area is required.

Excellent sow and boar observation is possible.

The slats reduce daily routine time.

Good control of sow feed regulation is possible.

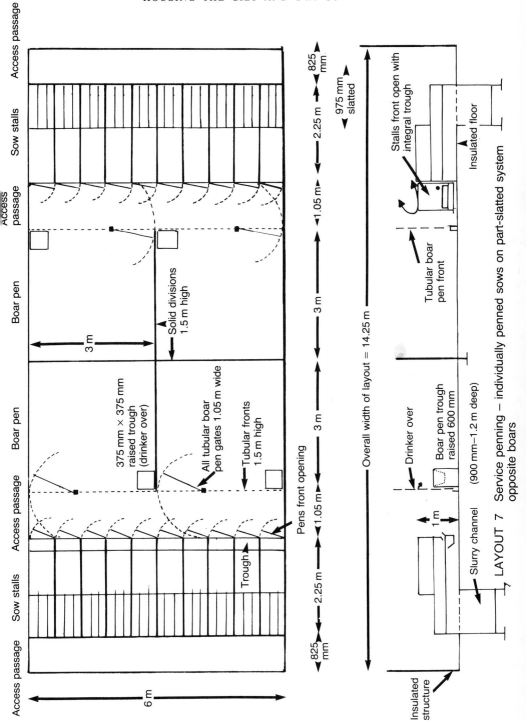

LAYOUT 7   Service penning – individually penned sows on part-slatted system opposite boars

It is a more convenient use of space if one sow is slow to show oestrus or returns to service.

Disadvantages

Individually housed sows may take longer to show signs of oestrus.

There is no nose-to-nose contact between sows and boars.

The boar pens require manual cleaning.

SOWS IN PREGNANCY

A wide range of factors have to be considered when making the choice of dry sow accommodation. There are two basic types of housing systems: group housing and individual housing.

Plate 2 Sow yards with tractor scraped yard between bedded lying kennel and individual sow feeders.

There are alternative layouts for each type. Five group and three individual layouts are described and discussed in this section.

The relative merits of these two systems are as follows:

GROUP HOUSING

Advantages

Utility buildings may be used.

Bedding can be used.

Free movement of sows favoured by welfarists.

Contact allows oestrus or illness to be more rapidly observed.

Disadvantages

The 'out of step' sow is expensive to house in pens designed for a number of pigs.

Bullying can be a problem even if sows are individually fed.

Bedding can be extremely expensive to transport, handle and store.

A concrete apron is required outside the building on which to store or from which to collect the muck.

INDIVIDUAL HOUSING

Advantages

Good control over feed intakes.

Lower cost.

Conception and litter size may be better.

Allows better use of penning when sows get 'out of step' with their group mates.

Disadvantages

Needs a specialist structure.

Oestrus stimulus and observation is less easy, particularly in a large herd.

As most systems rely upon mechanical control of temperature some fail-safe provision is advisable.

Slurry-based systems, on which individual systems might well be based, create a need for storage or special disposal considerations.

The trend towards an increasingly large scale, specialist unit and the need to maximise pig performance are most likely to convince the farmer that, despite criticisms levelled against restriction of sow movement, the individual housing methods do offer a greater opportunity to achieve operational and financial efficiency. However, the growing influence of the welfarist and the need to find uses for straw in certain areas may be sufficient to convince some that a lower level of output may be tolerable. It is all a matter of personal preference and judgement.

LAYOUT 8

Kennel and Yard (with Bedding) and Individual Feeders (Six Sows per Pen)

Important Considerations

The sows should be allocated 1.3 m² (14 sq ft) of lying area each, but in other respects this layout is as for Layout 1 described under gilt housing with the same operational methods and merits and disadvantages.

It is important to note that:

A well-laid 150 mm (6 in) thick concrete manure pad (possibly reinforced) is required onto which to scrape the muck.

Up to 800 kg of straw per sow place per year may be required with the appropriate straw storage provisions.

Automatic feeding is difficult.

All group housing buildings for sows demand draught-proof end doors to be fitted and that either space-boarded or plastic-clad external cladding above 1.5 m (5 ft) from passage level be more enclosed during cold weather if conditions demand it.

Plate 3　Sow cubicle layout with tractor scraped yard. The width of the three individual lying/feeding cubicles is the same as the passage width. (See layout 9).

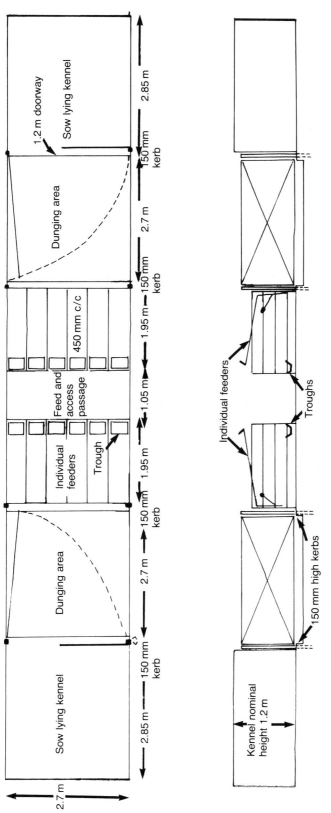

LAYOUT 8 Dry sow penning – bedded kennel and yard layout with individual feeders (6 sows per pen)

LAYOUT 9

Cubicles (Three Sows per Pen)

Important Considerations

The cubicle layout was developed as a means of reducing the costs associated with group housing whilst increasing the degree of individual protection for the more timid sow.

It is necessary to provide good fitting doors and gates and to draughtproof the cubicle.

As described in Layout 2 it is necessary to profile the rear kerb to retain bedding. This is an essential requirement for the operation of this layout.

The system works very much better on the layout shown, because an access passageway between the dunging areas eases sow movement and observation. The original cubicle layout incorporated a single central feed passage and no access or stock inspection other than via the dunging passage, which cannot be recommended except for the economy of space.

A normal feature of design is for a depressed trough, normally with a hard tile base. This trough, formed in the cubicle floor profile, is normally only 50 mm (2 in) deep and feed is dropped manually or automatically into it from above.

Water provision is normally from a drinker fixed to the gate, which closes into a centre cubicle when sows are shut into the lying area. This is connected via a flexible hose to a water supply clipped just inside the cubicle lid to reduce the risk of freezing. Alternatively, the drinker may be fixed above the kerb on the passage side of the dunging area with care taken to protect it from tractor scrapers and to insulate the main supply from frost.

Merits of the System

There is a 3 per cent saving in floor area compared to the yard and feeder layout.

There is better control of bullying with the individual lying cubicle.

The sows may be automatically fed.

Disadvantages

Some sows persist in vulva-biting their pen mates that are stood in the cubicle. To overcome this a double-hinged gate or plastic flaps may be used. These may additionally help in temperature control.

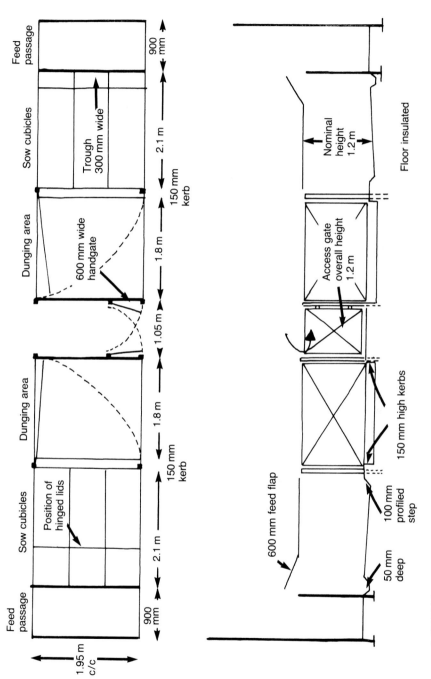

LAYOUT 9 Dry sow penning – sow cubicles (3 sows per pen)

LAYOUT 10

Deep Straw Yards (Four Sows per Pen)

Important Considerations

In order to permit the use of big bales of straw a layout with gate posts with a number of fixing points which allow muck to build up for only occasional tractor removal can be developed.

The building shell should be insulated, even with generous straw provision.

The individual feeders are set up on a raised area so that up to 650 mm (2 ft 2 in) of muck can build up. Drinkers would be sited at the side of the ramped area.

Operating the System

Individual feeders are used to enclose the sows for occasional bedding with big bales which are separated or rolled out in the bedded area.

Sows are best admitted and removed from the pen via the front-opening feeders into the centre access pasage.

Merits of the System

Less time is required for routine bedding and cleaning.

Big bales of straw can be utilised.

Stock observation is good.

Costs of walling/kennelling are lower, but the structure may be more expensive due to the need to insulate.

Disadvantages

Sows tend to become very dirty in hot weather due to wallowing.

It is difficult to automate feeding because of the use of front-opening feeders.

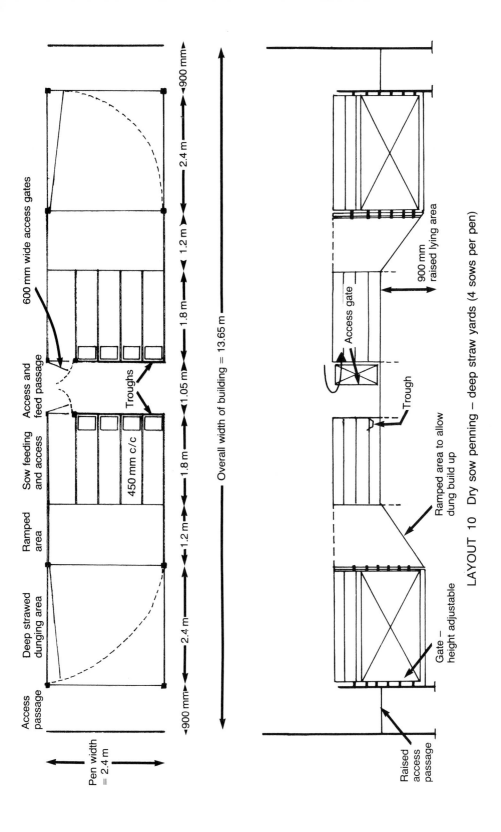

600 mm wide access gates

Access and feed passage

Troughs

Sow feeding and access

450 mm c/c

Ramped area

Deep strawed dunging area

Access passage

900 mm • 2.4 m • 1.2 m • 1.8 m • 1.05 m • 1.8 m • 1.2 m • 2.4 m • 900 mm

Overall width of building = 13.65 m

Pen width = 2.4 m

900 mm raised lying area

Access gate

Trough

Ramped area to allow dung build up

Gate – height adjustable

Raised access passage

LAYOUT 10 Dry sow penning – deep straw yards (4 sows per pen)

LAYOUT 11

Voluntary Stalls with Slatted Loafing Area (Three Sows per Pen)

Important Considerations

This layout requires precise control of house temperature due to the absence of bedding and because the sows lie individually.

The stalls have a double gate arrangement so that when the sow enters the stall she pushes a front gate up which causes the rear one to close, keeping out her pen mates. The reverse arises when she backs out into the loafing area. Slat dimensions are discussed under individual sow housing later in this chapter.

If a fixed trough is used it should be set into the floor to allow sows to get more comfortably over it into the access passageway.

A variation is to extend the slats under the stalls by up to 1 m (3 ft 3 in) to allow their use as stalls.

Operating the System

Feeding and inspection is from the centre passageway.

A manual locking device allows sows to be enclosed in the stall for removal, pregnancy testing, etc.

The ability to enclose sows allows bullying, etc. to be reduced.

Merits of the System

Less space required per sow compared to a bedded system.

Observation of the sows is excellent.

It can be used as an individual arrangement for sick or timid sows.

Disadvantages

Automatic feeding is not possible due to use of front-opening pens.

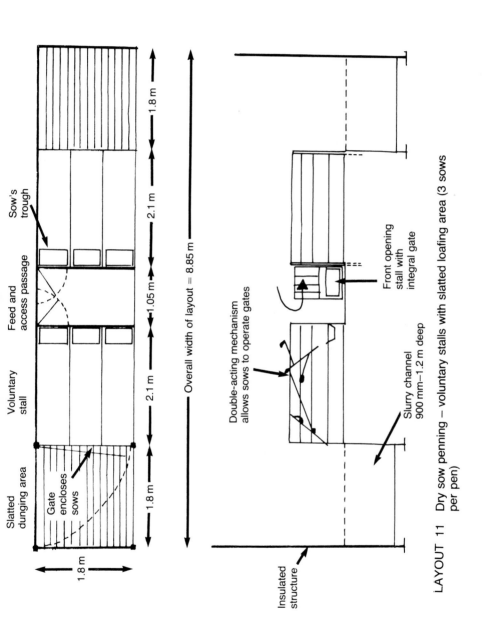

LAYOUT 11 Dry sow penning – voluntary stalls with slatted loafing area (3 sows per pen)

Layout 12

Loose Housing With Computer-controlled Automatic Feeding (Sixteen Sows per Yard)

Important Considerations

Using the space provisions as for Layouts 8, 9 and 10, a simple kennel or insulated but bedded structure, this method seeks to provide an accurate method of automatically feeding and monitoring sows.

A tubular feed stall, or station, is placed to give access from a pair of yards.

In two twelve-hour shifts, or possibly more frequently as experience is gained with the system, the dividing handgate is altered allowing a second group of sows access.

At a remote terminal the daily feed allocation for each sow is programmed and, at the conclusion of the group feeding period, a printout shows how much of the daily allocation has been consumed.

Experience is currently limited with the system but it appears that up to forty sows, in two groups, can obtain the appropriate amount of feed from one feed station each day. Trials are under way to determine the number of visits sows should pay to the feed station in order to gain their full daily requirements but observations suggest sows prefer to gain their full daily intake at a single visit.

A double-acting gate on the station prevents another sow entering the feed stall. A responder situated in the collar around the sow's neck activates the 'interrogator' in the feed station to allow the sow to be identified and the correct feed delivered into a trough.

Experience shows that a 2 m × 2 m (6 ft 6 in × 6 ft 6 in) area behind the feed stall is necessary and the overall length of the station itself, including feed store, should be 2.5 m (8 ft 4 in) and a 2 m (6 ft 6 in) width should be allowed for the feed stall outside the main pen itself. Early evidence suggests that the greater the area behind the feed station the less aggression is likely to occur.

Operating the System

It is normal for sows to be grouped when they are more than six weeks pregnant.

The feed allocation is programmed according to sow condition and can be adjusted as necessary.

Merits of the System

Excellent control of feed intakes is achieved.

A simple building layout can be used and a smaller area per sow is required than in a yard and feeder layout.

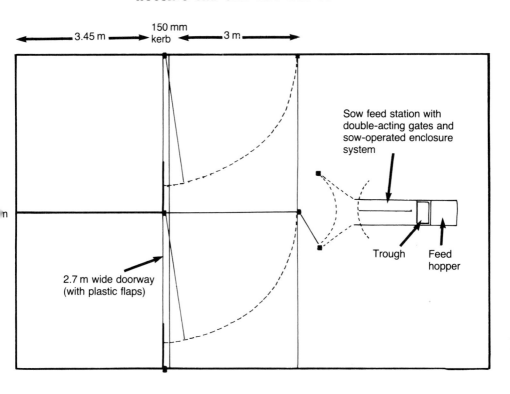

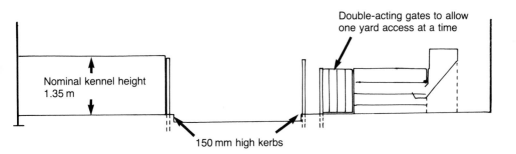

LAYOUT 12 Dry sow penning – loose housing with computer-controlled automatic feeding (16 sows per yard)

Users claim considerable sow contentment.

The amount of labour is reduced.

The computer can be linked to a sow recording system.

Disadvantages

The computer controlled system is costly.

There are risks from electricity failure.

It is necessary to ensure operators physically inspect the stock.

Some timid sows are reluctant to use the feed stall, although most settle within a few days. Sows have to be 'trained' to use the system initially.

The occasional 'rogue' sow will 'steal' another sow's feed and it is necessary to be ready to remove this sow and the very timid one to another system.

There is a risk of fighting when sows are mixed and the group size is best suited to the larger herd. One feed station is suitable for approximately a hundred sows in the herd.

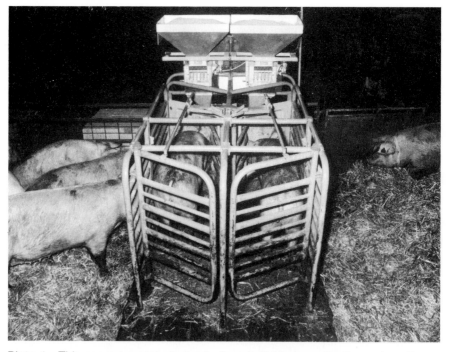

Plate 4   This computer controlled unit allows individual sow feeding and monitors those which do not eat fully. It is used in conjunction with a loose housed system.

LAYOUT 13

Individual Housing with Bedded or Slatted Loafing Area

Important Considerations

This layout seeks to balance the benefits of individual sow penning with the opportunity for pigs to move around and thus counter welfare criticisms.

Central, kennelled lying cubicles are arranged on a zigzag layout and sows reverse into a loafing area (the width of two pens) via a double-acting door hung on the rear of the kennel.

Mechanical scraping of loafing area is only possible with specialist, narrow wheelbase tractor or hand-steered scraper.

Operating the System

The sows are fed from a trough in the yard.

Access is gained from the outer feed passages via a gate in the loafing area front.

Bedding is normally used in the kennels.

Merits of the System

Excellent control of the sows is possible.

It can be used as a service layout in conjunction with the adjacent boar pens.

Disadvantages

The capital cost is high.

It is difficult to observe sows if they are in the lying area.

It is not easy to remove sows from the lying cubicle.

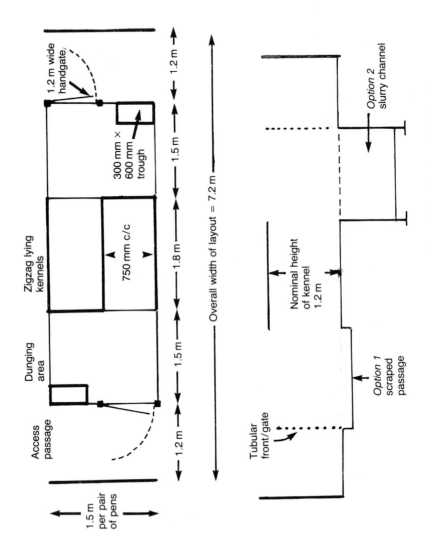

LAYOUT 13   Dry sow penning – individual penning with bedded or slatted loafing area

LAYOUT 14

Sow Stalls—Part-slatted

Important Considerations

A critical feature of this individual sow system is the control of temperature and ventilation. Many problems associated with both stall and tether systems stem from poor temperature and draught control, also from the maintenance of lower critical temperature at the expense of humidity and odours. Recirculation ventilation systems are only partly successful. They may help to maintain temperature but odour levels may be unacceptable. The only answer is the provision of supplementary heat to allow for greater air exchange.

This layout cannot be recommended for solid floors and hand cleaning except on a small scale due to the difficulties in scraping away manure. If a slurry channel cannot be used, a raised, cantilevered design is advised, possibly with a portable ramp to ease sow entry and removal from the stall. Muck is then removed by scraper from below the raised slatted area.

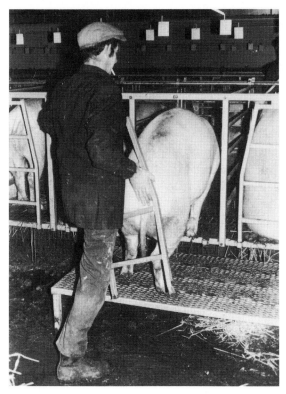

Plate 5 This cantilevered sow stall layout allows some bedding to be used as the muck is scraped from below the raised pen.

The length of the stall depends on the trough arrangement. Where a fixed trough with the lip above floor level is used the standing should be 2.4 m (8 ft) long. If a depressed trough is used the sow can lie more comfortably and so a shorter 2.1 m (7 ft) standing may be used.

Stalls should be positioned at no less than 600 mm (2 ft) between divisions, preferably at 635 mm (2 ft 1½ in) centres for sow comfort.

The bottom rail of the stall should be approximately 175 mm (7 in) above floor level close to the trough and 225–235 mm (9–9½ in) at the rear of the standing.

The stalls should be at least 1 m (3 ft 3 in) high.

The stall division over the front 600 mm (1 ft) above the trough should be of solid steel plate (not mesh) to prevent bullying, and this plate should be extended to form a trough divider to within 12–25 mm (½–1 in) of the trough profile.

To ease sow inspection and pregnancy testing the rear 750–900 mm (2 ft 6 in–3 ft) of the stall may be cut away to a height of only 600 mm (2 ft).

Tubular rear gates cannot be recommended as they tend to discomfort sows. Chains suspended at 250 mm and 400 mm (10 in and 16 in) from the floor offer a comfortable retainer but allow dung and urine to soil the passage. A solid metal or timber rear gate is advised with care taken to ensure that it fits in line with the edge of the slat opening. This prevents dung build-up on the slats which may cause bacterial build-up leading to infection of the sows.

When used for pregnant sows, automatic or semi-automatic feed dispensers may be fixed to the stall fronts and in large sow stall houses they reduce stress at feeding time.

The solid portion of the pen floor should be insulated and laid to have a 25 mm (1 in) fall to the slats. Care must be taken to set the legs of the stall into the sub-floor to give a rigid fixing. These legs should be of solid steel to prevent corrosion.

The slats themselves are vital in terms of sow comfort. A well-made concrete slat is probably the most suitable for the size of the sow's foot and durability. The slat itself should have a moulded edge to increase sow comfort and the slat should be 125 mm (5 in) wide with a 25 mm (1 in) gap between the slats. The slatted area of the standing is normally adequate if it extends 1 m (3 ft 3 in) from the rear gate of the stall.

There are alternative means of watering; by float valve maintaining a constant level or by hand operated or time-clock controlled valve into a trough laid essentially level. It is important that the water discharge point is no further than 12 m (40 ft) from any sow to ensure satisfactory water flow.

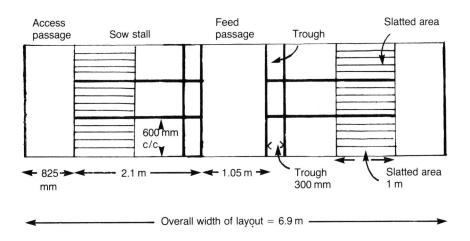

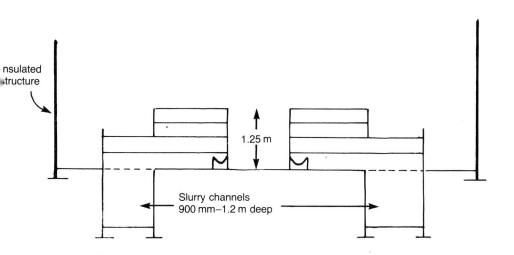

LAYOUT 14   Dry sow penning – part-slatted sow stalls

Operating the System

Once-daily feeding is normally practised and sows are inspected immediately following feeding.

A manual watering system is normally operated 30–60 minutes after feeding and may be repeated at intervals during the day.

Checking for oestrus is best made by passing a boar at the head of the sow with the stockman observing the sow from behind.

Sows are normally penned in sequence of anticipated farrowing to simplify management.

Merits of the System

It has a low floor space requirement.

There is excellent control of each sow and bullying is eliminated.

The animals remain very clean.

Disadvantages

Observation of return to oestrus is not always easy.

Temperature/odour control is often a compromise when ambient temperatures fall below 10°C (50°F).

LAYOUT 15

Sow Tethers (Neck or Girth Secured), Part-slatted or Solid Floors

Important Considerations
Temperature and ventilation control observations, slat dimensions, trough and watering comments are of equal importance to those made in Layout 14.

The tether layout may be used over slats as described in Layout 14, and it is normal to totally slat the area between each row of sows when sited back-to-back.

If a bedded layout is preferred a depressed gulley behind the sow is normally used. It is important to avoid excessive floor slopes or standings which are too short, otherwise prolapse may occur.

Divisions should be at 600 mm (2 ft) centres and 1 m (3 ft 3 in) nominal height.

Length of divisions differ according to whether a neck or girth tether is used:
Neck Tether. The division should be 1.05 m (3 ft 6 in) from the front edge of the trough with the tether secured to the lower rail of the division.

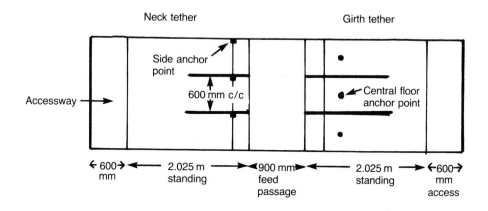

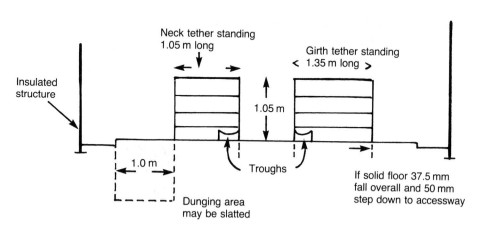

LAYOUT 15   Dry sow penning – sow tethers on part slats or solid floors (neck or girth)

Girth Tether. The division should be 1.35 m (4 ft 6 in) from the front of the
trough with the tether anchor point set centrally in the floor of the
standing 150 × 225 mm (6–9 in) behind the edge of the trough.

Provision of solid head dividers, trough plates and solid legs of divisions as for
Layout 14.

The length of the floor of the standing (which should be insulated) should be
no less than 2.025 m (6 ft 9 in) from the front of the trough, with a fall of
30–35 mm (approximately 1½ in) from the trough to a 50 mm (2 in) high step.

If solid floors are used a 600–750 mm (2 ft–2 ft 6 in) passage between the step
of the slurry gully is advised and it is essential to either install drains or
drainage gulleys to remove excessive urine from the house.

Operating the System

As Layout 14.

Merits of the System

It has probably the lowest capital outlay of any dry sow system.

It gives the opportunity to use bedding.

It is convenient to check sows for pregnancy.

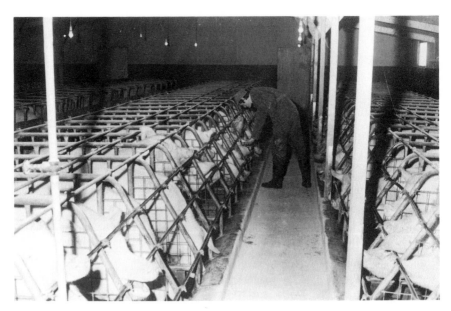

Plate 6   A typical sow stall layout; sow tethers would be similar. The rear 1 m of the
pen would be slatted.

Disadvantages

It is difficult to secure the sows.

It is inconvenient if the sows have to be removed for service.

The noise of the chain movement is distracting during staff operation, particularly when pregnancy detection is being undertaken.

Cost Comparison

A comparison of the costs of the various layouts described is almost impossible due to the variations in qualities of materials used, the degree to which farm labour is employed, site conditions, etc. However, provided there is no restriction on materials and that a comparative cost of labour is applied, the guidelines given in table 7.1 may be used. It is fair to point out that there is a greater opportunity for local or farm labour to be used where the general purpose building and kennel-type layout is used than with a fully insulated structure. Table 7.1 assumes that all work would be conducted by hired

Table 7.1   A comparison of the costs of the housing systems in chapter 7

Layout No.	Description of System	Narrative	Relative cost per sow Housed
1	Bedded kennel, yard with individual feeders	Gilt housing	130
2a	Voluntary stall, bedded yard	(including share	115
2b	Cubicle with bedded yard	of boar pen cost	115
3	Part-slatted pen	in each case)	100
4	Voluntary stall with bedded yard	Service area	125
5	Alternate sow and boar yards, bedded	(including share	115
6	Yarded sows with boars opposite, bedded	of boar costs	130
7	Stalls opposite boars, slurry based	in each case)	100
8	Bedded sow yards with individual feeders	In-pig sows	150
9	Cubicles with bedded yards		140
10	Deep straw yards		140
11	Voluntary stalls with slats		120
12	Yards with computer-controlled feed system	Includes feeder costs	160
13	Zigzag lying area with bedded yards		180
14a	Part-slatted stalls	In-pig sows	120
14b	Stalls over raised cantilevered slats		130
15a	Part-slatted tethers		110
15b	Tethers with bedded, solid floors		100

labour. In broad terms, building costs may be broken down into three near-equal components: site and baseworks; building shell; pen fittings and the provision of services. These ratios do vary according to the system chosen but constitute a useful guideline.

Table 7.1 is shown as shared cost per animal housed *not* per square area of floorspace or cubic area of airspace. Automatic feeding is not included in the comparisons unless stated.

Chapter 8

HOUSING THE BOAR

THE SPECIFIC needs of the boar have frequently been neglected. All too often he is allocated an area of the dry sow quarters and little regard is given to his special needs.

Due to their greater mature size and the need to maintain them in an active state so as to stimulate reproductive activity in sows, it is preferable to provide boar penning which gives them the opportunity to move around in their pens and to do so in close proximity to sows.

The penning of boars within a house does tend to exaggerate temperature control problems in housing in a temperate climate. The boar is allocated a relatively large area of a building which relies upon other animals in the house to keep to an adequate degree of warmth. It is wrong to design boar penning on the assumption that the boar is a much more robust animal because, unless special provision is made, both he, and those who share the same airspace, may spend prolonged periods of time being too cold.

Critical Considerations

- The house must be sufficiently insulated and the ventilation controlled so that house/pen temperature stays above the lower critical temperature of 19°C if bedding is used, or 22°C if it is not.
- It is vital to remember that because boars do not have the opportunity to lie in contact with other pigs, they have greater difficulty in offsetting low temperatures.
- The provision of kennelling is essential in those housing layouts such as yards and cubicles where there is an uninsulated structure and large building airspace to facilitate mechanical cleaning. It may be necessary to go without some tubular divisions for good sow:boar contact to create a more comfortable temperature for the boar.
- The provision of solid divisions between pens is advisable so that the boar does have some opportunity to lie quietly at times to avoid harassment from other boars as well as very active and excited sows.
- Boars are also susceptible to higher temperatures in which they tend to become lethargic workers and, as proved by documented evidence from around the world, exposure to high temperatures for prolonged periods of

time drastically reduces fertility and this reduction remains so for many weeks. In practice, every effort must be made to maintain boars in temperatures below 25°C, and where ambient conditions make this difficult the use of water cooling should be considered.

- An adult boar should be provided with an area of 7.5 m² (80 sq ft) for lying, feeding and dunging alone. If the boar pen is also a combined service area, the size should be extended to 10 m² (108 sq ft) with no division less than 2.4 m (8 ft) in length.
- The height of the pen divisions should be no less than 1.5 m (5 ft) and between boars these should be solid to prevent intimidation. Where access to the sows is provided, vertical tubular divisions are advised to prevent damage to limbs when boars serve. These vertical divisions should be at approximately 125 mm (5 in) centres. It is advisable to construct the tubular walls on a kerb or dwarf wall to increase boar comfort and to reduce spillage of bedding into access passages.
- The importance of providing a good pen floor is vital to all classes of pigs but almost particularly so for boars. The importance of a slip-proof surface at mating time is apparent in the improvement in results which arise from a change from slippery service area floors. To provide such a floor is easier when service and living pens are separate as the opportunity to wet the floor is reduced. The following points are stressed:

    1. Boars should not be expected to serve in a pen with a slatted floor because uncertain foot adhesion can occur giving depressed results.
    2. Boars should not be expected to serve in an area which is mechanically scraped as this eventually becomes increasingly slippery.
    3. If a combined living and service pen is provided the floor must be well drained and bedding should be used. Such pens should be cleaned and the floors checked prior to service routine being commenced.
    4. A separate service pen may have a floor of peat, crushed chalk or sand to provide a non-slip surface.
    5. In a combined pen if the boar is expected to serve in the lying area it is necessary to ensure that the ceiling line is at least 2.1 m (7 ft) high. All ceilings should be well protected from potential damage by boars.
    6. No pen floor should have a slope greater than 1:50 as such slopes tend to create trauma of the hoof particularly when a boar is mounted. This can create some difficulties in a combined living and service pen where bedding is used, particularly as pigs tend to move their bedding much more than other species and this impairs drainage.

- A service area should preferably have completely enclosed sides to prevent distraction of both boar and sow by others. The service area should have no obstructions upon which a pig may damage itself and some find it an advantage to 'round out' pen corners to prevent sows standing in an inconvenient position. As mentioned, no service pen division should be less than 2.4 m (8 ft) in length and 2.4 m × 2.4 m (8 ft × 8 ft) may itself be a satisfactory service pen size. Anything larger will certainly be relatively inconvenient for gilt services.

- In a combined service and living pen it is essential that careful consideration is given to the positioning of the boar's trough. For greatest convenience this would be sited in a corner of the pen adjacent to the access passage. The trough itself should be raised from the floor so that it is not possible for the boar, or a sow, to slip in it. A satisfactory design for boar pens is to provide a trough 300 mm × 300 mm (1 ft × 1 ft) with the lip of the trough 450 mm (1 ft 6 in) above the pen floor. It helps to keep pen floors dry and clean if a bite action drinker is situated a further 200–250 mm (8–10 in) above the trough.

Careful choice of drinker type for a boar is necessary because, almost more than any other class of pig, they tend to hold individual foibles on the type of drinker that they are willing to use. A stockman needs to be ready to change a drinker if a boar shows an unwillingness to use a particular type.

Boar Pen Layout

Chapter 7 provides suggestions for boar pen layout in Layouts 4–7. Layout 6 shows a separate living and service pen configuration; however, greater detail of a separate service pen is shown in Layout 16.

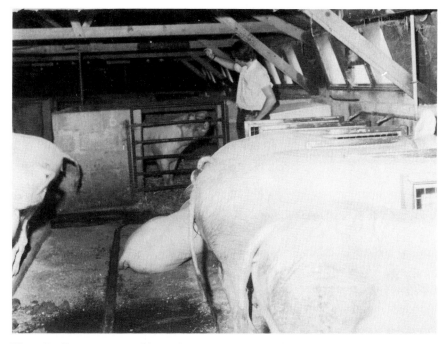

Plate 7 Boar contact with newly weaned sows and those due to be checked 3 weeks later is vital to a good service layout.

LAYOUT 16

The Service Pen

Important Considerations

Pen sides should be no less than 2.4 m (8 ft).

The pen corners should be rounded out by the use of 300 mm × 300 mm (1 ft × 1 ft) corner diagonals.

The pen door should have a quick-release catch for stockman safety and it should be possible to open it in both directions.

If it is inconvenient to use a hard-core floor, or if an existing concrete floor is too slippery, two alternatives may be considered:

—To use a proprietary rubber mat in the service pen to enhance floor grip.
—To abrade the existing concrete using dilute hydrochloric acid followed by a thorough washing of the surface or the use of an abrading machine. These would roughen the floor without exposing the aggregate.

Operating the System

Before service routine commences the floors should be checked to ensure that they are not wet and slippery.

The sow would be placed in the service pen first and the boar then moved into the pen. The reverse routine is advised after checking/service.

Merit of Separate Service Pen

No compromise on shape or size of pen is necessary.

The opportunity to provide a slip-proof floor is increased.

A better living area for the boar can be provided.

Disadvantages

A greater cost may be involved.

The service routine takes longer as service pens have to be shared by a number of boars, although this may improve stockman observation of service.

Some boars are distracted when moved to a pen in which another boar has been recently penned and may have served and this may increase the safety hazard.

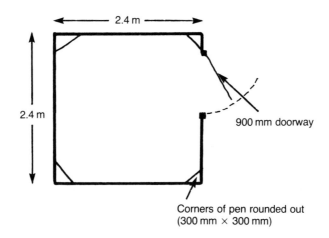

900 mm doorway

Corners of pen rounded out
(300 mm × 300 mm)

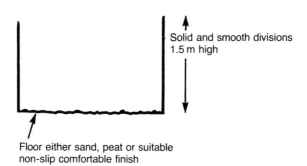

Solid and smooth divisions
1.5 m high

Floor either sand, peat or suitable
non-slip comfortable finish

LAYOUT 16 The service pen

Chapter 9

THE FARROWING PEN

THE GROWTH of modern production methodology makes the achievement of high levels of output essential and more difficult to achieve through high levels of stockmanship alone due to the large number of pigs being tended. This means that greater attention needs to be provided to farrowing pen design to help reduce piglet losses.

Unfortunately, there are a number of conflicting issues which arise from the penning together of the smallest and the largest pigs on the farm. Acute differences arise from:

- Different temperature requirements and tolerances.
- Disparate spatial requirements.

Although sows are capable of farrowing and rearing piglets to a good level of efficiency in a loose housing system, indoors or outdoors, these relative levels of output are not suited to housed, specialist systems on a commercial basis. It is essential to provide conditions giving priority to piglet survival while at the same time making the operators' job as convenient as possible. There are many vital design considerations regardless of the actual pen layout chosen, but no layout which does not incorporate a farrowing crate can be considered for a sow with piglets under one month of age.

GENERAL CONSIDERATIONS

Temperature Requirements

Sow comfort is more likely to be achieved if the farrowing room temperature is kept in the range of 16–21°C. If temperatures exceed this, distress may occur, particularly at farrowing time, and appetite will be depressed resulting in an excessive loss in condition. Temperature requirements for the baby piglet, particularly the new born, are totally different. A piglet subject to a temperature range suited to the sow is almost certain to die or, at best, thrive less well. For the first 24–48 hours of life piglets require a temperature of 30°C. Up to three weeks of age the temperature should not fall below 24°C in the area where the piglets lie and sleep.

There is little hope of a reconciliation of these temperature differences unless an enclosed, separate creep area is provided. Pens with no separate area are bound to create conditions where sows are too warm, piglets too cold or—worse—both.

From a practical viewpoint it is desirable that the general room temperature is maintained at a minimum of 21°C until all sows have farrowed, with supplementary heat provided near to the sow's tail for the first 24 hours after farrowing. After a few days it should be safe to allow room temperature to reduce but the creep temperature should be maintained at the higher level.

Considerations of precise, yet disparate, temperature levels accentuate the need to provide:

- A well-insulated structure.
- Complete draught elimination, with particular attention being paid to slurry channels, if used.
- Good ventilation control.
- A properly designed creep for piglets.

Creep Design

The following points of detail should be applied:

Floor Area
This must be large enough to allow all the litter to lie in a comfortable position up to the age of weaning or removal. For a system where pigs are weaned in the first month of life an area of 0.8 m² (8.5 sq ft) should be provided.

Height
The height of the creep should be such to allow suspended heaters to be safely fitted. A height of between 450 mm–600 mm (1 ft 6 in–2 ft) is recommended.

Temperature Control
Within a well-insulated structure it is not easy to justify insulation of the creep, although if the creep borders an external wall of the building or a conversion is being used, then the wall and lid of the creep should be given additional insulation. There are two approaches to reduce heat loss into the general area of the farrowing room:

- To use small 225 mm (9 in) wide popholes, which can then be easily shuttered to enclose piglets in the creep for routine tasks.
- To use wider creep entrances and to create a curtain of plastic strips cut to just brush the floor surface.

Inspection
The creep area itself should be easily hinged to allow good piglet inspection. The main danger with the covered creep is that operators are unable to observe an ailing piglet easily. However, the advantages in temperature control and running costs outweigh the operational inconvenience.

Heating

Creep heating by provision of manual or automatic heater regulation is advisable. The aim is to attract piglets to the creep, so that they lie without huddling, yet are never so hot that they are reluctant to lie in the covered area. As a guide, creep temperature should be 30°C at birth, reducing to 26°C at weaning, if this occurs at under one month of age. There are several main forms of creep heating:

Suspended infra-red lamps are cheap to install, attract piglets to the creep but must be within an enclosed area because they can cause air currents, pulling cold air over piglets not directly under the lamp itself. The main drawbacks are fire risk and breakage of the lamp. The bulkiness of the lamp means that the area of the creep cover in which the lamp is positioned cannot conveniently be hinged.

Suspended electric bar heaters can normally be fixed to the underside of the creep cover and so may be less likely to be knocked down, and are less of an encumbrance to lid adjustment. They are more expensive than an infra-red lamp and need supplementary attraction/inspection lighting.

Floor pads are sealed units on which piglets lie. Thermostatic control is needed to prevent excessive heat build-up and supplementary lighting to attract piglets is necessary.

Heat panels fit on the underside of the creep cover. Made in factory-finished sizes with fixing points, these may replace the creep cover entirely but their lower heat output means that they may be less effective unless the creep is properly enclosed. Again, attraction lighting is needed.

Radiators may be used where an economic source of heated water is available with a thermostatically controlled motorised valve to control the temperature. The radiator itself may form a creep/pen division. They are more expensive to install and attraction lighting is needed.

Under-floor heating comprising coiled hot water pipes may be used. Alternatively, electric elements carefully laid upon a sand:cement screed which in turn is laid over an extruded polyurethane insulant can be used. The electric elements demand very careful laying to avoid damage to the cables and extended curing time to prevent cracking of the concrete. To avoid the latter problem it may be preferable to lay the cables in a sand bed and to cover this with an asbestos cement floor panel. An alternative approach is to use hot water passed through pipes set into the floor or a low voltage electric current passed through wire hawser. Control of electric floor heaters is by thermostat with a remote sensor bulb connected by capillary tube mounted on the floor. It is vital that heavy-duty industrial electric cable is used for farrowing house purposes. The normal loading requirement is 150 W per creep in a covered creep and 300 W if uncovered. Maintenance costs and potential problems with leakage/failure have deterred general adoption of underfloor heating, although it can be very successful and little risk of fire exists. Separate attraction lighting would be required.

Attraction/inspection lighting is recommended. A badly lit covered creep is much less likely to attract piglets and will make the stockman's task less convenient. It is advisable to have supply cables outside the creeps and to use heavy-duty light bulbs as the creep temperature creates brittleness in these fittings.

Pen Floors

These have been a source of problems in pig units for many years. The balance between animal comfort, foot adhesion and wear are rarely satisfactorily resolved. Where a floor is less abrasive, to reduce joint and teat damage to the new-born, it is often too slippery for the sow to raise herself comfortably, and such floors may be subject to damage by pressure washers. There is still no more satisfactory floor surface than a well-mixed, properly cured 2:1 or 3:1 sand:cement screed with wood float finish and the coarseness removed by steel trowel. New additives, including polyesters, may prove to make satisfactory floors. Recent adoption of asphalt-based tiles laid on a prepared concrete surface has contributed to animal comfort, although in the initial wearing areas, by the sow's trough and her front feet area, this surface would require renewal at 12–15 month intervals, thus satisfying all requirements except long life.

Slats

The quest to increase farrowing house hygiene and sow and piglet comfort has led to the utilisation of a slatted pen portion and, in some cases total slatting. Just as the disparity in requirements between the sow and litter lead to the need to compromise in temperature provision and solid pen floors, the same dilemma occurs with slats, perhaps to an even greater degree. Because slats are used to increase cleanliness of the stock, it is vital that they allow good passage of manure and, in general, void areas below 40 per cent do *not* meet this requirement.

The dimension, shape and edge of the slat must be compatible with the hoof size of the new-born piglet and so the hole should be no wider than 10 mm (less than ½ in) and the edges rounded. However, with smooth-surface slatting materials such as polypropylene, aluminium or perforated panels, a degree of raised but rounded foot grips would be needed to ensure foot adhesion for the sow without damaging the joints of the piglets. The dimension of the slat itself must be comfortable for piglets to walk and lie upon and be easy to clean and disinfect between batches. Although initial cost of slats may influence the choice of material used, it is vital to consider potential length of life of the slatting material used. It must also be borne in mind that many cheaper materials may require a greater degree of support thus reducing or eliminating any cost advantage. Table 9.1 summarises the merits of slats for use in farrowing pens.

Table 9.1   Outline appraisal of slatting materials in farrowing pens

Material	Price comparison	Support required	Piglet comfort	Sow comfort	Cleanliness	Length of Life
Welded mesh	100	Yes add 50%	Fair	Good	Excellent	Less than 5 years
Expanded mesh	110	Yes add 50%	Not good	Teat damage likely	Good	Less than 5 years
Woven wire	240	Yes add 25%	Fair	Satisfactory	Very good	Excellent
Perforated metal*	172	No	Not good	Slippery	Fair	5–8 years
Aluminium alloy	356	Yes add 25%	Good	Slippery	Quite good	May be poor
Galvanised bars*	450	No	Fair	Excellent	Good	Excellent
Cast iron	510	No	Excellent	Excellent	Excellent	Excellent
Plastic coated mesh	370	Yes add 50%	Excellent	Slippery	Fair	Around 5 years
Polypropylene	420	Yes add 50%	Excellent	May be slippery	Fair	Good
Concrete	170	No	Good	Good	Unacceptable	Excellent

*Denotes galvanised finish.

The Farrowing Crate

The farrowing crate was developed to reduce piglet losses by controlling sow movement. With the advent of early hybrid sows small crates were designed, but the development of the typical crossbred sow means that mature body size is now greater and crate dimensions should be adjusted to suit that fact. As in many other respects in the farrowing house there are a wide range of requirements which must be catered for. For example, a mature sow can be fifty per cent longer and taller than a first litter gilt and have a girth up to twice the size. Although operators do not care for adjustments on crates, it is not possible to meet the special needs of such variable pen occupants satisfactorily without the ability to modify the basic crate dimensions. Certain requirements are now clear and before farrowing crates are purchased, checks to establish how they compare to the points below should be made.

Crate Length
The larger sows will measure approximately 2 m (6 ft 6 in) in length, so the length that the sow can actually use for lying must allow for a rump guard which should be 225 mm (9 in) from the tail gate to allow free piglet movement, sufficient space at birth and the sow's trough. Therefore, any crate less than 2.4 m (8 ft) overall length will be too short for longer sows. Crate length adjustment should be made by moving trough/crate front assembly rather than altering the rump guard to ensure that pen cleanliness is maintained.

Crate Width
Fewer piglets are laid on if the sow is forced to lie down slowly by first kneeling, lying on her belly and then adjusting to the prone position for suckling. To make a sow lie in this way the crate width between horizontal bars should be around 500 mm (1 ft 8 in). This is too restricted for

comfortable lying, so the bottom rail must either be bowed or raised or fitted with prongs, adjusted mechanically or hydraulically controlled. The width for comfortable lying for suckling is about 735 m (2 ft 5½ in). Whatever bottom rail configuration is used this dimension should be provided. A fixed or adjustable bottom rail must be no higher than 250 mm (10 in) above pen floor level and this itself may be restrictive in allowing adequate suckling space if a large sow with a full udder is in the pen. Neither can it be lower due to the risk of gilts becoming trapped. Thus to give greater all round comfort for sow and litter a raised rail 350 mm (1 ft 2 in) with prongs with tips 100 mm (4 in) from the floor near the sow's head may be preferable. Prongs may be graded to shorter length to the rear of the crate to increase sow comfort and suckling access but, near the sow's head, they should be spaced no wider than 200 mm (8 in) to prevent the sow's head becoming trapped.

Crate Height
An overall height of 1 m (3 ft 3 in) should comfortably accommodate a sow even when her back is arched for feeding or urinating.

The Trough
The necessity to feed suckling sows generously is widely acknowledged, so the trough should have good capacity to allow up to 7 kg of feed to be offered without spillage. A trough should be deep and wide with an anti-splash lip on the sow's edge of the trough. The trough should have a curved base or be tilted towards the sow to make feeding more comfortable. An anti-splash plate above the trough allows feed to be placed in the trough without spillage. Ideally, there should be a 250 mm (6 in) clearance below the trough to allow the sow to lie with her head beneath it and utilise the full crate length.

Drinkers
There are two preferred drinking arrangements. One is to have a separate bowl drinker beside the trough and raised to prevent piglets falling into it. This is operable by all sows but is prone to wetting floors and needs hand cleaning. The alternative is to site a nose-operated drinker over the trough 150 mm (6 in) from the trough side and 200 m (8 in) from its base, so that the sow operates the drinker and then drinks from the trough. Whilst the latter choice is cheaper to install and reduces floor wetness, not all sows operate such a drinker fully and the wetting of feed means that it more quickly becomes stale if the sow does not rapidly clear up all the feed offered.

The Crate Rear
The need for a rump guard has been mentioned. This should be rounded and vertical to prevent vulva damage to the sow. The rump guard may be attached to the rear gate of the crate although if the crate is to be hand cleaned it may be more convenient to use a separate rump guard and rear gate assembly. There is an advantage in having the rear 750 mm (2 ft 6 in) of the crate cut away down to the bottom rail to allow better access to the sow in case the need to assist farrowing arises.

The Farrowing Pen

Crate length and creep dimensions begin to dictate pen size and shape. The crate is to be 2.4 m (8 ft) long and, if a forward creep is used, a further 525 mm (1 ft 9 in) is added to the pen length. Either side of the crate, which will have an overall width of 600 mm (2 ft), side escape ways at least 300 mm (1 ft), and preferably 450 mm (1 ft 6 in) must be added. This gives a pen area of approximately 4.4 m² (47 sq ft).

Pen Divisions

The pen divisions should be at least 400 mm (1 ft 4 in) high to prevent piglets jumping from pen to pen. It is vital that materials which are easily cleaned and disinfected are used for this purpose.

Piglet Drinkers

Debate exists over the need to provide supplementary watering facilities for piglets weaned at less than one month of age. If installed, a drinker should be positioned near to the rear of the pen to prevent floor wetting.

Piglet Farrowing Boxes

This is an attachment fixed at the rear of the farrowing crate over which a heater is suspended. The intention is for piglets to enter this warm area and be unable to get out by use of a 250 mm (10 in) step. This reduces the risk of chilling and savaging until the stockman is on hand. The box is normally 500 mm (1 ft 8 in) wide and 600 m (2 ft) long, constructed of easily cleaned panelling.

Passages

The large amount of piglet handling demands a generous central passage of 1.2 m (4 ft) regardless of the layout used. If muck is to be removed by hand, a passage no less than 1.05 m (3 ft 6 in) should be used. A rear passage in a part-slatted system used for stock access and inspection need only be 825 mm (2 ft 9 in) wide. If pens are raised to allow muck to be withdrawn from below slats by hand, the gap under the slats should be 375 mm (1 ft 3 in) and the passage width no less than 1.2 m (4 ft) to allow convenient muck removal. It is normal to slope rear passages to the slurry channel or drain to speed up floor drying.

Farrowing Pen Layouts

Farrowing house layouts may be judged against the criteria mentioned in this chapter. It is vital to ensure that the house/pen layout not only affords good piglet protection and sow and piglet comfort but allows ease of feeding sows and litters, inspection of piglets and removal of dung which are all important considerations. It is possible to satisfy the pigs' needs with the minimum of compromise and give the operator conditions under which he or she may be more successful.

Ideally, farrowing accommodation is laid on so that no more than one week's farrowings are placed in one room. However, there is an upper limit at which convenient between-batch cleaning becomes difficult to arrange. Twelve farrowing pens per room appears a convenient maximum number and this suits approximately a 300-sow herd depending on age at weaning.

The layouts discussed below all assume that the appropriate design features are provided. The dung removal systems shown are interchangeable, given adjustment to passage width. Although the operational advantage of using a three-passage room layout is obvious, it must be remembered that this increases floor space requirements by approximately 1.3 m² (14.5 sq ft) per pen and this is not always easy to justify except, possibly, in the larger herd.

Plate 8   A bedded farrowing crate layout with forward, boxed creep. Note the lamp at the rear of the pen for farrowing.

LAYOUT 17

Farrowing Pen with Forward Creep and Part-slatted Pens

Important Considerations

Pen dimensions must conform to standards given. Slats are 1 m (3 ft 3 in) long with a floor slope of 25 mm (1 in) from the creep to the slat.

The piglet drinker (if used) is sited over the slat.

If a suspended creep heater is used it should be positioned centrally in the creep to prevent piglets fouling one side of the creep box.

Power and water supplies should be clipped to the creep lid to ease cleaning and aid stock inspection from the centre feed passage.

Operating the System

Sows are fed over the forward creep with a solid splash plate directing the feed into the trough.

Slurry is removed between batches.

Piglet creep feed is offered from the rear of the pen.

A lamp is suspended over the slats at farrowing time.

Merits of the System

Feeding the sows is quick, easy and well suited to large herds.

Pen cleanliness is usually excellent as pigs normally dung over the slatted area.

Generous passageways ease pig movement and handling.

There is little contact with the outer wall of the house.

Piglets may be enclosed in the creep to ease conduct of routine stock tasks.

Disadvantages

There is a relatively large area per pen.

Rear end inspection involves extra walking for the operator.

Creep feeding demands additional walking by the operator.

The covered forward creep is not always readily used or 'found' by the litter and piglets may have to be enclosed for a short period to encourage them to use the creep.

Inspection of the sow over the creep from the front of the pen is not easy.

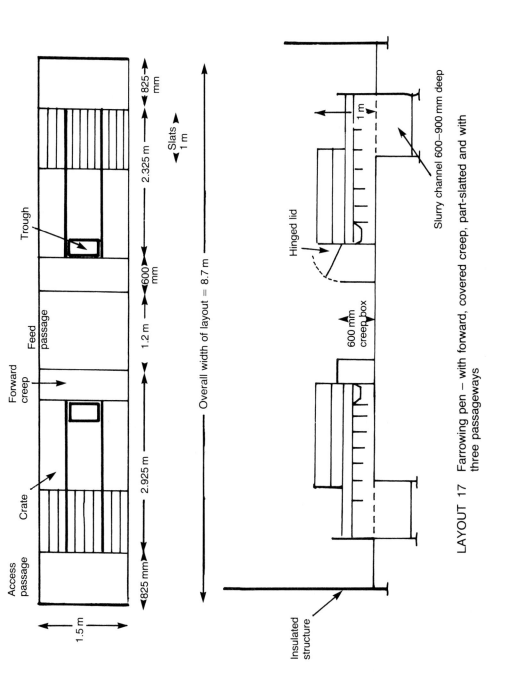

LAYOUT 17 Farrowing pen – with forward, covered creep, part-slatted and with three passageways

LAYOUT 18

Side-covered Creep Layout with Part-slatted Pens

Important Considerations

The crates should be offset to give wider area on one side of the pen. 600 mm (2 ft) is recommended but a shorter pen length is needed as no forward creep is used.

The creeps are arranged in pairs so that services may be shared.

The side of the crate against the creeps should be sheeted to prevent the sow damaging the creep fittings.

Operating the System

The sows are fed from the centre passage with a splash plate directing the feed into the trough.

Piglets are creep fed from the rear passage, i.e. outside the covered creep.

The piglet drinker is sited over the slatted area.

The heater may be shared between the pair of creeps.

Plate 9   This part-slatted layout uses a covered, offset side/front creep.

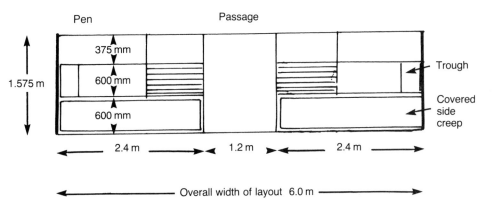

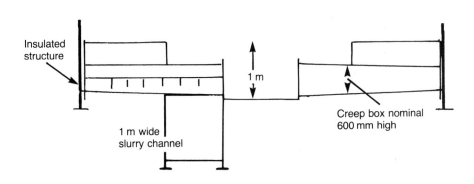

LAYOUT 18   Farrowing pens with side-covered creep and part-slatted pens with
single passage

Merits of the System

A slightly smaller pen is required.

Sow inspection is easier than in Layout 17.

It is possible to remove the sow from crate through the front of the pen.

Disadvantages

All side creeps, except Layout 20, have the disadvantage that the piglets dung in the front corner of the pen on the side opposite to the creep. This corner should be rounded out or covered with a grill which drains to the rear slats.

LAYOUT 19

Farrowing Pens with Solid-floor Dunging, Side-covered Creeps and Single Passage

Important Considerations

The passage must be wide enough to allow manure to be removed from under the raised, slatted area.

Because some creeps will be against an outside wall, good insulation is needed.

Operating the System

The sows are fed by stepping up into the pen from the centre passage.

Creep feed is added from the centre passage on the widest side of the pen.

Manure is removed daily, by hand, from below the slatted area.

The sows can easily step up and down a height of 375 mm (1 ft 3 in).

Merits of the System

It saves passage area thus reducing the cubic area needed to be warmed, as the eaves height of the building may also be slightly reduced.

A single passage means all feeding is from the same passage.

The daily removal of muck keeps down odour levels.

It is useful for conversions where no slurry channel exists.

Disadvantages

The height up into the pen is inconvenient for feeding sows.

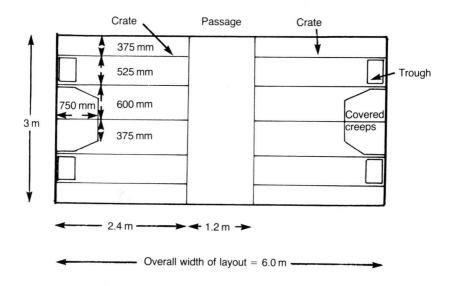

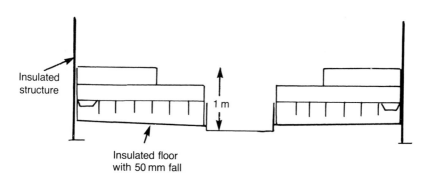

LAYOUT 19 Farrowing pens with solid-floor dunging, side-covered creeps and
          single passage

Any step-in arrangement increases the risk of the spread of disease.

The offset creep encourages pigs to dung on the opposite side of the pen at the sow's head.

Hand removal of manure is time consuming.

It is not easy to 'reverse' the sow out of the crate.

It is more expensive.

LAYOUT 20

Farrowing Pens with Inserted Creep on Part-slatted Layout and Three Passages

Important Considerations

The creep is constructed to jut into the adjacent pen so as to save space and prevent piglets dunging in the fore area of the pen.

The creep must be properly protected against damage from the sow.

Operating the System

The sows are fed from the central passage.

The creep feed is given from the rear passage.

Manure is removed between batches.

Merits of the System

The pen floors are cleaner as piglets dung at the rear of the pen.

It is possible to create a larger creep without extending the area occupied by the pen.

The sows can be removed from the front of the pen.

Disadvantages

An irregular creep shape is less easy to construct.

At each end of a row of crates there is a degree of wasted space but this could be used as a bedding store or for something similar.

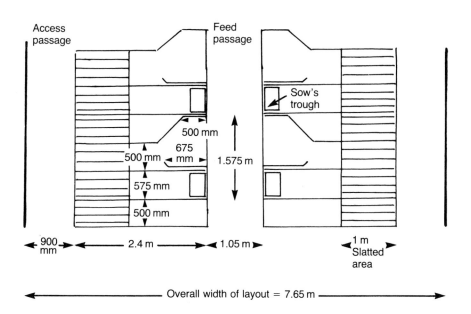

Access passage

Feed passage

Sow's trough

500 mm

675 mm

500 mm

1.575 m

575 mm

500 mm

900 mm

2.4 m

1.05 m

1 m
Slatted area

Overall width of layout = 7.65 m

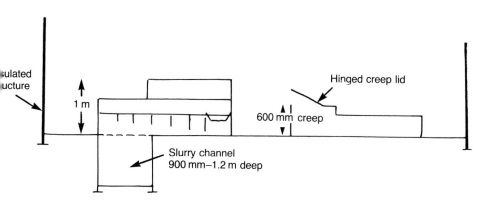

Insulated structure

Hinged creep lid

1 m

600 mm creep

Slurry channel
900 mm–1.2 m deep

LAYOUT 20 Farrowing pens with inserted creep on part-slatted layout and three passages

LAYOUT 21

Herringbone Layout with Covered Creep and One Central Passage

Important Considerations

Where room width is limiting, it allows two rows of crates to be fitted.

The pen area itself makes no particular saving of floor space.

Operating the System

The sows are installed via the small, triangular step-in accessway from which manure may also be removed.

The sows are fed by leaning over the creep from the step-in area.

Creep access and feeding are also from the main passage.

It is possible to arrange the layout so that the creep cover and the pen side are removed and the sows are released from a side-pivoting crate.

Merits of the System

It suits a narrow structure.

Piglet access and observation are good.

It deters piglets from dunging at the head of the sow.

Disadvantages

The shape of the pen is not convenient to construct.

N.B. A variant on this layout is to use a rectangular pen with a crate set diagonally. The widest area nearest to the sow's head is used as a covered creep zone.

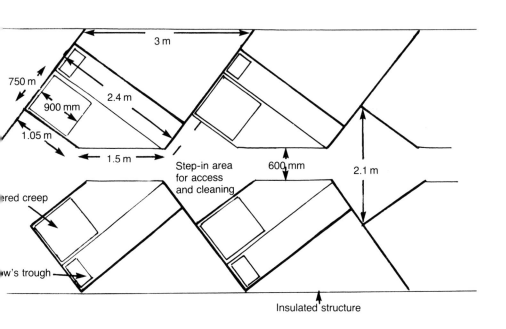

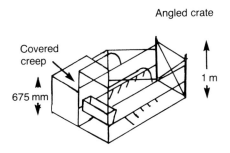

Angled crate

LAYOUT 21   Farrowing pens – on herringbone layout with covered creep and
central passage

Layout 22

Covered Side/Forward Creep and Step-in Accessway from Central Passage

Important Considerations

From the centre passage the stockman can step into a narrow 300–400 mm (1 ft–1 ft 4 in) wide accessway to feed and inspect the creep area. This accessway finishes close to the creep boxes which are arranged in pairs between two crates.

Operating the System

The sows are fed and the creeps inspected from the step-in accessway. Creep feeding is also conducted on the same side of the pen.

Merits of the System

It gives excellent access with a regular shape of pen yet only one main house passage.

No stepping into pens is required for routine tasks.

Disadvantages

The irregular creep box shape is not easy to make and is only large enough for litters weaned up to 3 weeks of age.

The piglets will dung near the head of the sow opposite to the creep box if the pen corner is not 'rounded out' or a slotted panel inserted down that side of the pen.

Slightly more material is needed for the pen surround than on some of the other layouts.

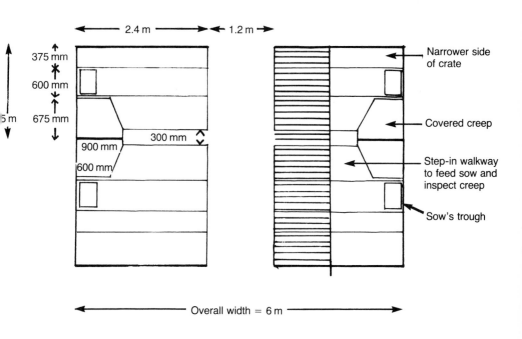

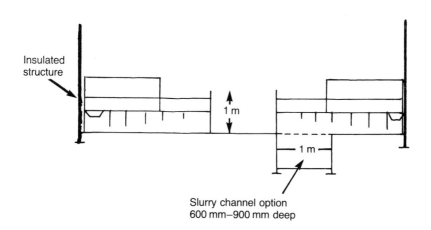

LAYOUT 22 Farrowing pen with covered side/forward creep and step-in accessway from central passage

FOLLOW-ON PENNING FOR SOWS AND LITTERS

Although it may be a little cheaper to house sows and litters from two weeks onwards in simpler follow-on penning, it is not normally convenient to do so due to the doubled time required for pressure-cleaning, the additional pig movement involved and the possibility that hand cleaning would be required. As only a relatively small cost saving would be made, few producers weaning at under one month of age would choose to use a follow-on approach. Units where later weaning is practised or where it is planned to delay weaning a few litters containing sub-standard piglets are those where follow-on sow and litter penning are likely to be used.

Individual layouts offer better control of the sow and litter but are more expensive to construct and necessitate hand cleaning. The simple monopitch layout (Layout 23) is used as a farrowing layout by some operators, but piglet access is so poor that it has become less popular.

LAYOUT 23

Monopitch, Individual Sow and Litter Follow-on Pen (The Solari)

Important Considerations

The pen width is arranged to allow reasonable piglet protection from over-lying and a creep large enough to accommodate litters to 15+ kg.

The pen height at the creep is normally 1 m (3 ft 3 in) and 2 m (6 ft 6 in) at the front of the pens, and a heater is provided in the creep.

Ventilation is controlled by top-hinged flaps which close to the 1.2 m (4 ft) high pen front over which sows are fed in a fixed trough.

Operating the System

The sows are fed over the pen front from outside the pen.

Creep feed is given by entering the pen or via an airtight slide in a low rear wall.

The manure is hand removed from the pen front and bedding is added via the same route.

The sow and litter are moved from the specialist farrowing pen when the litter is a minimum of ten days of age. Piglets may remain in the same pen for a few days after weaning.

The temperature is controlled by flap adjustment and the addition or removal of bedding.

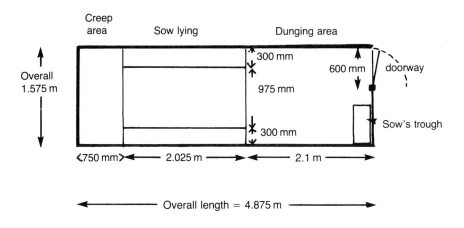

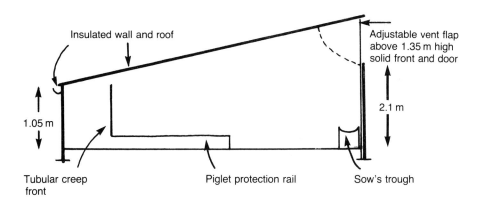

LAYOUT 23   Follow-on pen – monopitch pen for single sow and litter

Merits of the System

It is a simple structure with a low running cost.

Young litters may be moved to this system.

Separate pens simplify between-batch cleaning.

Disadvantages

It is relatively inconvenient for sow and creep feeding.

The labour requirement may be higher.

LAYOUT 24

Multi-suckling Yards for Sows and Litters on Bedded Principle

Important Considerations

A floor area of 7 m² (75 sq ft) per sow and litter (excluding passage and external accessways) is required.

The height of the building front should be 3 m (10 ft) to allow between-batch manure removal by tractor loader. Manure is normally allowed to build up between batches and generous amounts of straw are added.

It is preferable to arrange a 225 m (9 in) deep step mid point in the sow's lying area to demarcate the lying area and keep it drier and cleaner. The lower dunging area can then be given an exaggerated fall to encourage drainage to an outer, dished gully.

The monopitch roof profile would drop 1.5 m (5 ft) over the creep area which would have a 1.05 m (3 ft 6 in) high lid hinged to give access for inspection and feeding. Heat pads are the safest form of heating in such a layout.

An excavated rear passage provides good headroom for the operator.

Approximately 500 kg of bedding for four sows and litters per month of occupation would be required.

Operating the System

Sows are fed via ad-lib feeders on a raised floor area which is filled manually or via an overhead auger or in sow feeders stood on a concrete apron in front of the monopitch structure.

Creep feeding is via the rear passage into small hoppers.

Bedding is added as necessary and manure removed between batches.

Ventilation is controlled by light flaps hinged from a front roof member and raised or lowered depending upon conditions.

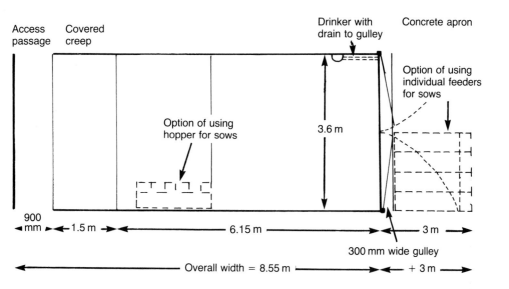

Access passage

Covered creep

Drinker with drain to gulley

Concrete apron

Option of using individual feeders for sows

Option of using hopper for sows

3.6 m

900 mm

1.5 m

6.15 m

3 m

300 mm wide gulley

Overall width = 8.55 m

+ 3 m

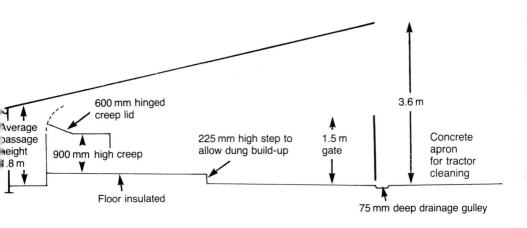

600 mm hinged creep lid

Average passage height 1.8 m

900 mm high creep

225 mm high step to allow dung build-up

1.5 m gate

3.6 m

Concrete apron for tractor cleaning

Floor insulated

75 mm deep drainage gulley

LAYOUT 24   Follow-on pen – multi-suckling yard system for four sows and litters on bedded principle

Litters should be three weeks old (eighteen days minimum) before being mixed into this system.

It is often used for litters up to twelve weeks of age as a weaner pool after removal of the sows.

Merits of the System

It is a useful system for a range of pigs.

Daily routine is reduced compared to individual follow-on pens.

Disadvantages

It tends to lead to a wide disparity of piglet weights and sow condition.

It cannot be safely used for piglets under eighteen days as it lacks flexibility.

It has a high straw requirement.

Cost Comparisons

Table 9.2 summarises the relative costs of layouts described in this chapter.

Plate 10　A bedded multi-suckling system with a large rear, heated creep (lid raised). Sows feed communally or via the outside individual feeders.

Table 9.2   Relative costs of layouts described in chapter 9

Layout	Description	Narrative	Relative costs per farrowing pen
17	Forward creep and part-slatted		125
17b	Forward creep with solid floor	Needs wider passage	115
18	Side creep and part-slatted		115
18b	Side creep with solid floor	Needs wider passage	100
19	Single passage, side creep, hand cleaned		100
20	Inserted side/front creep and part-slatted		110
21	Herringbone, side creep, part-slatted		115
21b	Herringbone, side creep with hand cleaning		110
22	Side/front creep single passageway		105
			Relative cost sow and litter
23	Monopitch rearing pen		120
24	Multi-suckling		100

The Sow Washing Crate

Many producers acknowledge the merits of sow washing prior to placing in, and even after removal from, the farrowing pen. To allow this task to be efficiently conducted with minimum stress to the sow, a sow crate should be sited over slats or a drain in a warm area. The crate itself should be no less than 2 m (6 ft 6 in) long and 1 m (3 ft 3 in) high with vertical tubular rails positioned 250 mm (10 in) apart. The top of the crate should be movable to permit good access to all parts of the sow.

There should be no less than 750 mm (2 ft 6 in) clearance all round the washing crate to give good access for the operator. Hot and cold water points should be positioned adjacent to the crate along with suitable shelves to hold shampoo, skin dressings, etc.

Chapter 10

WEANER PENNING

A RANGE of weaner pen designs is commonly used. The earlier that pigs are weaned the more precise are the demands for temperature control. The point at which less sophisticated housing might be adopted is largely determined by the weight of the pig at weaning. One mistake which can be made is to use the *average* weight of the pigs as the arbiter of design. It is the *minimum* weight of the pigs to be weaned that should be used as the basis of such a judgement, as it is the smallest piglet which is the most susceptible to unsuitable and variable temperature levels.

Because of the direct relationship between temperature control and the weight of the piglets at weaning, the type of accommodation used should be considered on the basis of minimum pig weight at weaning, and this chapter is divided accordingly.

General Considerations

- Pigs weaned at under 6 kg require a carefully controlled temperature which is difficult to provide in manually controlled layouts, e.g. kennels.

- Good husbandry will ensure that very small piglets are kept alive even in less good housing systems, but the growth efficiency of such pigs will be far below that possible under conditions with better temperature control.

- In systems where mechanical control of temperature is used it is important that the ventilation rate is sufficient to expel the build-up of gases and humidity. Important factors are:

  1. Insulation of the building structure.
  2. Cubic capacity of the building.
  3. Acceptance that a certain amount of supplementary heating will be needed.

- In addition to the need to provide the appropriate floor space, it is also important to consider the number of pigs penned together. There are three important principles in this respect which should always be given consideration:

1. If some degree of controlled feeding is practised it is important to match the number of pigs to the feeding space available.
2. The smaller pigs in a batch of weaners should always be separated from the rest.
3. It is useful to plan to sub-divide any group of weaners into at least five pens, and also never to exceed twenty pigs per pen.

- The trough and hopper design should be appropriate for the class of pig and feeding system to be practised. Piglets weaned under a month are often fed a restricted amount of feed for a period. This fact means that, even if a hopper is installed, it will be used as a trough for a period so this should be accounted for in design.

Troughs for Weaners
The use of a trough with a mesh retainer guard above improves pig inspection. This design is essential for tiered penning. The trough itself should be designed to minimise feed wastage. The key elements in prevention of wastage are:

1. The depth of the trough should be 85–100 mm (3½–4 in).
2. There should be a 12 mm (½ in) retainer 'lip' on each edge of the trough.
3. There should be trough dividers at 75 mm (3 in) centres to prevent piglets pushing feed from the trough.
4. The width of the trough at the top should be 125 mm (5 in).

Hoppers for Weaners
The hopper is normally used to form the pen front. It should be high enough to prevent pigs escaping from the pen. There is no merit in using a hopper with a capacity to hold more feed than can be consumed in one day. The atmosphere and type of feed used in weaner penning requires frequent additions of fresh feed. Therefore, a capacity of no more than 1 kg per pig penned need be provided for pigs up to eight weeks of age. The hopper should have:

1. A height of feed face of 75–100 mm (3–4 in).
2. A depth of feed face of 100–125 mm (4–5 in).
3. A return lip of 12 mm (½ in).
4. Trough dividers at 75 mm (3 in) centres.
5. A sloping feed face base to ensure that the limited quantities offered fall towards the pigs.
6. Openings at the feeder base should give controlled flow of feed without blockage or wastage.
7. A 50 mm × 50 mm (2 in × 2 in) step reduces the risk of pigs fouling the trough.

- The point at which pigs are transferred from the weaner accommodation to the grower penning is somewhat arbitrary and depends largely on farm circumstances and, to some extent, on the type of weaner accommodation used. However, it is normally quite reasonable to move piglets to less sophisticated buildings when the weight of the smallest pig is 12 kg. Higher level of growth efficiency may be achieved where pigs remain in the weaner penning up to 20 kg, as long as the spacial provisions suit that weight. These weight suggestions may well be totally different for pigs weaned with a minimum weight of 10 kg. Pigs in this category will normally be weaned at over one month of age and may therefore remain in their weaning accommodation until transfer to the finishing house. The problem with such a wide spread of age/weights in one penning system is that a considerable compromise on floor space allocation has to be made. Groups of weaners may have to be subdivided, and probably remixed, at some intermediary point. Neither of these choices is particularly attractive to a producer and so it may be more acceptable to move the pigs at certain intervals to pens large enough to accommodate the same group of animals as they grow.

- When planning a unit layout it may be quite acceptable to consider siting the weaner housing at some distance from the farrowing accommodation, as weaner piglets are most likely to be moved in a truck or mechanical transporter. It is later when they are larger and less easy to handle, that they need to be moved on foot. It is then that proximity to other buildings becomes more critical.

Plate 11   Tiered cages with mesh floors and dung collection trays are often used for weaning pigs below 5.5 kg in weight. Good environmental control is essential.

WEANING PIGS UNDER 6 KG LIVEWEIGHT

It is vital to establish the precise needs of this category of pigs. For the period up to 12 kg the temperature should be held at a constant 28°C with adequate provision for the removal of excess humidity and gases. In practice the temperature may be gradually reduced one week after weaning, but the house, insulation, ventilation and heating should be good enough to provide the above standards.

When weaning small piglets and providing penning to keep them in relatively small group sizes to give the desired performance levels, it is more convenient to provide totally slatted penning. The advantage in separating piglets from their dung at this stage to reduce digestive upsets is obvious. The provision of totally slatted penning precludes the use of bedding which is outside the general recommendations implicit in the Welfare Code Recommendations for pigs in the United Kingdom. It is no coincidence that the two layouts shown for weaning of piglets weighing less than 6 kg are both slatted *and* that the Welfare Code Recommendations specifically propose that no pig be weaned at earlier than three weeks of age. It is difficult to wean pigs in this weight category without the use of the type of layouts shown and, on economic grounds, these preclude the use of bedding. It is, however, the non-usage of bedding which creates the real welfare issue, not some unproven problems associated with a particular age at weaning. The stocking rate should be planned not to exceed 122 kg/m² (25 lb per sq ft).

It is normal to practise some degree of control over feeding during the immediate post-weaning stage. If this feeding method is to be used then it is vital to provide sufficient feeding space to allow the piglets penned to feed simultaneously. A trough/hopper face of 75 mm (3 in) per pig is required to permit this. If ad-lib or full feeding is to be practised, the feed face availability may be somewhat reduced to 50 mm (2 in) per pig provided that piglets of even size are penned together.

The siting of drinkers is also important. It may be more acceptable to provide a bowl-type drinker to ensure that all piglets use it. However, the fact that these drinkers become quickly fouled will probably lead to a nipple-type drinker being chosen. The nipple-type is used by the majority of pigs but is responsible for a greater degree of spillage. Nipple-type drinkers should be fitted so there is 225 mm (9 in) from the tip of the drinker to the pen floor level and positioned to be no less than 300 mm (1 ft) apart. Drinkers should not be positioned within 200 mm (9 in) of the pen corner due to the tendency of pigs to obstruct others drinking by dunging in the corners. A useful addition to the weaner house watering system is the provision of separate header tanks to allow medication to be administered when necessary. No more than ten pigs per drinker should be allowed.

LAYOUT 25

Tiered Slatted Penning for Pigs, 4.5–14 kg

Important Considerations

The pens are designed to hold eight piglets up to 15 kg and are normally 1.05 m (3 ft 6 in) long and 850 mm (2 ft 10 in) wide.

The pen height is normally 450 mm (1 ft 6 in) with a dung collecting tray sloping to a collecting gully between each pair of pens. The overall height of the pens is 1.95 m (6 ft 6 in) above floor level.

The pen divisions are of 40 mm × 40 mm (1¼ in × 1¼ in) mesh to prevent piglets becoming trapped yet allowing relatively unobstructed air movement.

Passages around the penning are no less than 600 mm (2 ft) wide.

Cross-flow ventilation is normally used with incoming air being pre-heated in a separate chamber prior to entry into the room. It is important to ensure that the ratio of room size to pre-heat chamber is not greater than 10:1. It is impossible to achieve correct environment control where a pre-heat chamber is shared between rooms. Ventilation is by use of variable speed fans.

Operating the System

Piglets are fed by hand two to three times daily with feed sprinkled in measured amounts along each trough.

Inspection of stock, checking of water and temperature are conducted at feeding times.

Manure is removed between batches.

Between-batch cleaning is carried out by removal of the pens, which are normally on castors. The pen fronts, floors and dung trays are also removed and pressure-washed. The water supply to each set of pens is detachable.

Pigs are placed into and removed from the pens by hand. Small or ailing piglets are normally placed in the middle tier to permit easiest inspection.

Merits of the System

Only a small space is needed for the number of pigs housed.

Easily dismantled penning allows comprehensive between-batch cleaning and maintenance.

Good temperature control is possible.

Disadvantages

Inspection and feeding of the upper tiers is inconvenient.

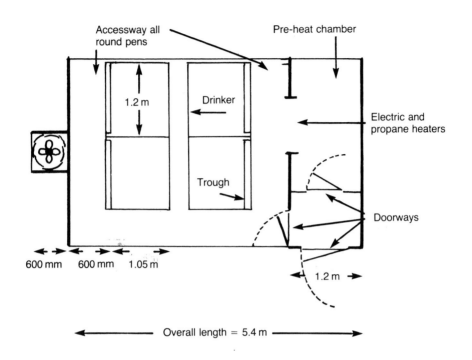

Accessway all round pens

Pre-heat chamber

Drinker

1.2 m

Trough

Electric and propane heaters

Doorways

600 mm   600 mm   1.05 m

1.2 m

Overall length = 5.4 m

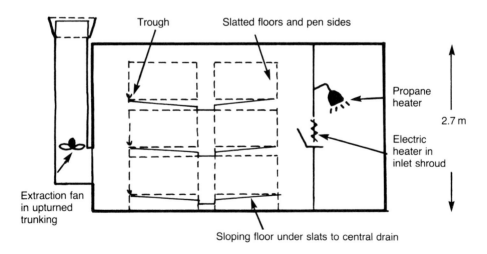

Trough

Slatted floors and pen sides

Propane heater

2.7 m

Electric heater in inlet shroud

Extraction fan in upturned trunking

Sloping floor under slats to central drain

LAYOUT 25   Tiered, slatted pens for 96 pigs from 4.5 kg to 14 kg

There is no opportunity for the pigs to be retained in this layout after 14 kg due to the size of the pen and the durability of the penning.

It is a difficult system for welfarists to accept.

LAYOUT 26

Flat Deck Penning (Ten Pigs per Pen)

Important Considerations

This layout of totally slatted penning with slurry chamber below is suitable for ten pigs up to 20 kg.

It has a pen division of 600–750 mm (2 ft–2 ft 6 in) high, which would be mesh or tubular divisions if cross-flow ventilation is used, but could be solid if a pressurised system is used.

The hoppers form the pen front.

The slurry channel should be 225–300 mm (9–12 in) deep, sufficient to hold slurry for up to five weeks of pig occupation and emptying between batches.

Hoppers are preferably positioned outside the length of the pens to allow the maximum floor area possible to be utilised per pig.

Operating the System

Pigs are normally penned by litter (with any very small pigs removed), according to size at weaning or according to size and sex at weaning. Small pigs may be placed in a raised pen sited above one of the pens with a trough and mesh pen front.

The room should be warmed to the required temperature before pigs are moved in.

Feeding is once or twice per day from the centre passage.

Inspection and routine checks are conducted at feeding time.

If the slurry channel is set below passage level the pigs may be moved to grower penning by foot.

The rooms are cleaned between batches as they normally hold no more than the number of pigs weaned in one week.

Merits of the System

It allows easy feeding of the pigs.

It permits excellent pig observation and inspection.

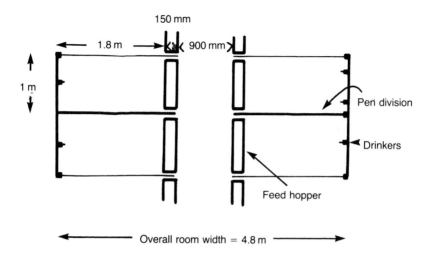

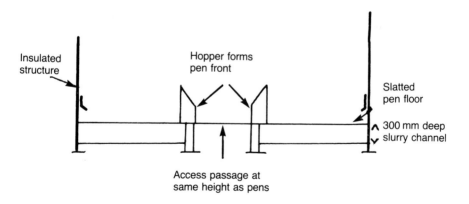

LAYOUT 26   Weaner penning – flat-deck penning for pigs from 5 kg upwards
             (10 pigs per pen)

Plate 12   A typical flat deck weaner layout with trough assembly used as a pen front.

The solid pen divisions make this a simple arrangement to thoroughly clean and disinfect; mesh or tubular panels are less easy.

Pig removal is easy.

The separation of a pig from its dung may help to control scours.

Disadvantages

Like all totally slatted systems there is a need to dock tails to minimise the risk of tail biting.

A greater area is needed than for the tiered system.

WEANING PIGS BETWEEN 6 AND 10 KG

In this category, pigs weaned at approximately 26–35 days, many would still choose a layout similar to Layout 26. However, because it is now assumed that the minimum weight is 6 kg, it is possible to consider less sophisticated systems. It would still be difficult to provide all the requirements listed below without supplementary heating.

The lying area temperature requirement should be a constant 26°C for the first week or so after weaning for the lighter pigs in this category. Larger piglets are more tolerant of a slightly lower temperature.

The assessment for floor space allocation is as in the previous section, i.e. 0.1 m² per pig (1 sq ft) lying area plus the same additional area for dunging and movement.

In layouts where at least part of the floor is solid it is possible to consider floor feeding. However, this is not recommended as the type of diet needed by this category of pig rapidly becomes unpalatable, and hopper feeding is more likely to ensure feed freshness and encourage better feed intake.

The provision of water will normally be by bite or nipple-type drinker sited in the dunging area of the pen. If pigs to be weaned are closer to the upper end of this category (10 kg) then the tip of the drinker may be somewhat higher from the floor, say 300 mm (1 ft).

Plate 13   These weaner kennels have underfloor heating and are sub-divided to give a warm nest. They may be used with slatted or bedded runs.

Layout 27

Heated Weaner Kennel (Fifteen Pigs per Pen)

Important Considerations

The lying area must be large enough to allow the hopper to be positioned without contravening the minimum lying area requirement.

Either electrically heated kennel lids or underfloor heating (manually or thermostatically controlled) should be used to ensure the minimum temperature is maintained. Plastic flaps would be hung across the 225 mm (9 in) wide × 225 mm (9 in) high pophole.

The inside kennel height is normally 600 mm (2 ft).

Ventilation of the kennel is either by manual adjustment of the kennel lid or the use of ACNV (see Chapter 3) controlled by a sensor positioned inside the kennel which raises or lowers the lid depending upon the temperature.

For younger weaners it is preferable to site kennels of this type under a covered building. This assists in temperature control and allows inspection of the piglets to be made in comfort.

Operating the System

Piglets are grouped in the pens as described in Layout 26, primarily by size and/or sex.

Feeding is carried out from the passage adjacent to the kennel.

Inspection is carried out from both sides of the pen.

Adjustment of ventilation is either manual or automatic.

Pens would normally be cleaned between batches.

Slurry would be removed by normal methods if a channel is used; in bedded systems manure would be removed weekly or fortnightly.

Merits of the System

It is less sophisticated than the fully slatted layouts and may have lower running costs.

Bedding can be used.

It is flexible in terms of the weight at which pigs are removed.

Disadvantages

One operator will find piglet observation very difficult, as the pigs rush about when the kennel lid is lifted.

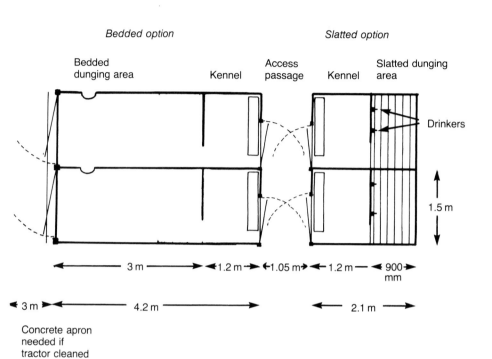

LAYOUT 27 Weaner penning – heated kennel layout (15 pigs per pen)

The use of an uninsulated structure increases the likelihood of the water supply freezing in the winter.

When the lid is lifted warm air is lost into the atmosphere.

LAYOUT 28

Portable, Part-slatted, Unheated Weaner Arks (Twenty Pigs per Pen)

Important Considerations

These are insulated, portable kennels sited over slurry channels or solid concrete with manure washed to a drain.

The kennel lid slides on bearers along the side of the lying area and is heavily insulated.

It is usual to subdivide the kennel at 900 mm (3 ft) from one wall and to place a feed hopper in the smaller lobby area.

There is a pophole normally 300 mm (1 ft) wide × 375 mm (1 ft 3 in) high which may be covered with a plastic flap.

Drinkers are sited over the slatted area.

As no supplementary heat is provided this layout also benefits from being sited under a covered structure.

Operating the System

This system is operated as for Layout 27, except that the system gives better results if larger piglets are placed in the pen.

Due to the larger number of pigs per pen this system is better suited to larger herds of 200+ sows.

Merits of the System

It requires relatively low capital and running costs.

Its portable nature is attractive, particularly for emergency penning or for a tenant farmer.

It is easily added to if herd size increases.

Disadvantages

Temperature control is poor; when the lid is opened warmed air is lost.

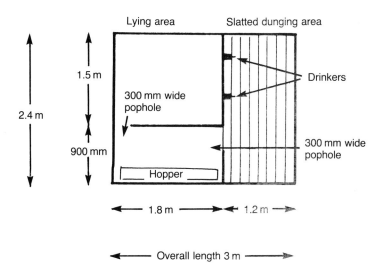

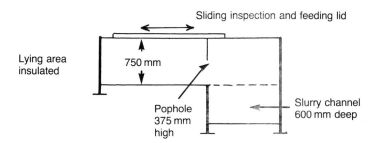

LAYOUT 28 Weaner penning – portable, part-slatted unheated arks (20 pigs per pen)

Inspection of the stock is difficult.

Water freezing may be a problem.

Its relatively lightweight construction may be easily damaged.

LAYOUT 29

The Verandah Layout

Important Considerations

It has a 2.4 m × 1.8 m (8 ft × 6 ft) lying kennel with 2.4 m × 1.2 m (8 ft × 4 ft) slatted run. The kennel may be 900 mm (3 ft) high with a hand-adjusted lid.

The pophole is normally 300 mm (1 ft) wide × 375 mm (1 ft 3 in) high.

The drinker is sited over the dunging area.

A removable cover is normally placed over the dunging area in the winter until the pigs reach 12+ kg.

The underside of the covered inspection area is insulated to prevent condensation. This area is ventilated by a raised ridge and an opening alongside the eaves just above the kennel lids giving an open area of approximately 1.8 m × 100 mm (6 ft × 4 in) per pen.

Operating the System

Pigs can be moved into the pen by foot, if desired, from the centre passage. A 900 mm (3 ft) door opens across a passage of the same width to ease pig movement.

The pigs are fed from the centre passage, preferably via a hopper.

Temperature is controlled by adjusting the kennel lid, and in hot weather by removing the cover over the dunging area.

The pigs are moved by blocking off the pophole and driving them out via the passage doorway.

The slurry channel is normally up to 1.2 m (4 ft) deep under the slats.

Pigs weaned at 10 kg may be housed at twenty to twenty-four per pen. If they are in the pens at over 30 kg, they should be reduced to eighteen per pen.

Merits of the System

It is very flexible as it can be used for a range of weights.

It has low operating and maintenance costs.

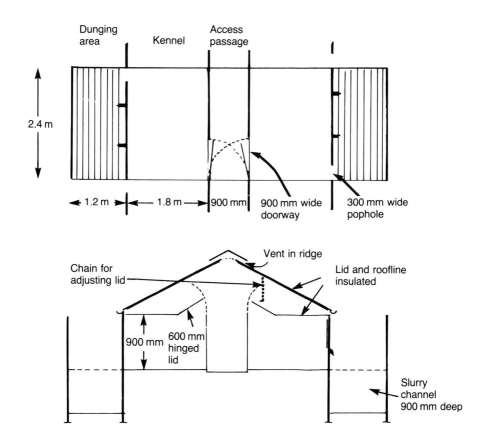

Original layout with uncovered dunging area

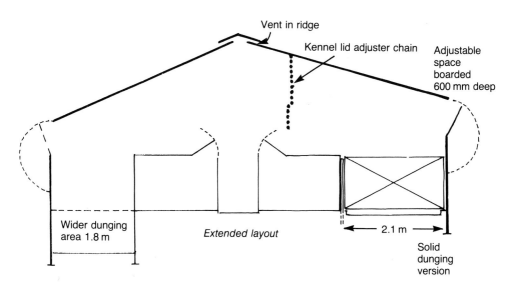

LAYOUT 29 Weaner penning – verandah pens, original and extended layouts

Plate 14   This verandah layout has a slatted outer run and a kennelled lying and feeding area.

Disadvantages

It is not suitable for very small pigs.

It has the usual problems associated with kennels of poor pig observation and water freezing.

The piglets may dung in the kennel if the pens are understocked in cold weather or overstocked in warm weather.

Cost Comparisons

These are summarised in table 10.1.

Table 10.1 Comparative costs of layouts described in chapter 10

Layout	Description	Narrative	Relative cost per piglet housed
25	Tiered penning	For pigs 5–14 kg	175
26	Flat deck penning	For pigs 5–20 kg	160
27	Heated kennels with solid floors	Minimum 6 kg at weaning Natural ventilation	130
27b	Heated kennels with solid floors	As above with ACNV control	145
27c	Heated kennels with slatted run	Natural ventilation	145
28	Weaner arks		100
29	Weaner verandah		120

Chapter 11

HOUSING THE GROWER

FOR MANY years the housing design for this category of pig received little attention. As a considerable proportion of pigs would be removed to a specialist finishing farm, it would be normal to place pigs of 20–45 kg in large groups in non-specialist buildings. Although some pigs are still penned in such layouts the need to maximise growth rates and reduce variation within batches of pigs has led to greater attention being focused on the grower.

There is a problem in design consideration when housing this intermediate stage between weaner and finisher. The pigs are increasing their weight extremely rapidly and so there comes a stage in their growth where a decision has to be made to either move them to a larger pen in the same group size or to subdivide the batch. In practice the decision is rather more complex than this. The heat output from the pigs themselves is normally used to maintain a suitable lying area temperature, so it is usually necessary to stock slightly higher numbers per pen initially to maintain an adequate LCT. This initial group size should not be so large so as to create a wide differential in growth within a pen. It should be compatible with the pen size in the finishing house, so that if growers are penned in larger groups than those in which they will eventually be finished, the group size should be able to be conveniently subdivided to suit the finishing house. There are three choices:

1. To maintain the same group size throughout the grower and finisher stages and to move pigs at regular intervals to larger pens.
2. To maintain the same group size throughout the grower phase, and to accept that the initial stocking capacity is wasteful of the floor area provided, and that there may even be a need, in severe weather, to use supplementary heat.
3. To subdivide pigs from a larger initial batch size within the grower house and, possibly, again on removal to the finisher accommodation.

The first of these three alternatives is probably preferable from the aspect of pig performance. It provides the opportunity for the changing space and ventilation requirements of the pig to be more closely met as it grows. The capital costs may be somewhat higher than a simple 'two-stage' system, because house subdivision may be necessary. The provision of rooms suited to a particular level of output is less conveniently extended to account for an expansion in production. Also the use of different diets is less easy to automate. However, as a principle of growing and finishing pig house design, this is the most likely to give excellent growth rates and allows between-batch cleaning to be practised which may increase performance, as will the subdivision of pigs from others in a different age/weight category. The disadvantage implicit in option 2 is eliminated because the stocking rate will be more compatible with the requirements of the pigs at various points in their growth.

The subdivision of the batch (option 3) is an easily extended system and requires two houses. Groups of differing size can be used between the grower and finisher phase. For example, pigs may be placed in groups of thirty pigs per pen in the grower house initially and subdivided into three pens of twenty pigs per pen at a given growth stage. It is feasible then that the three pens of twenty pigs may be subdivided further into two pens of ten pigs for finishing in the final house. From option 3, no matter how many dividing processes are used it can be seen that it saves very little operator's time and the remixing of pigs means that some fighting and stress may arise. In addition subdivision of pens is more complex where advantage is to be taken of split sex penning and feeding.

General Considerations

- Pigs in the weight range 20–45 kg would require a lying area temperature of between 14 and 24°C if kept on insulated concrete floors. This range would be increased to 18–25°C if the floors are totally slatted and reduced to 13–23°C if a straw-bedded house is used. These guidelines assume generous feed levels are adopted, i.e. pigs fed to appetite or ad-lib.
- Although feeding method and housing system may influence floor space allocation, it is useful to consider the floor area required for lying for pigs in the grower category:

    At 20 kg, 0.15 m² (1.6 sq ft) per pig
    At 30 kg, 0.2 m² (2.1 sq ft) per pig
    At 45 kg, 0.3 m² (3.2 sq ft) per pig.

    In totally slatted pens this floor space allocation would normally be increased by some 30 per cent to provide vice-free operation. In part-slatted pens it is advisable to double the allocation to ensure reduced pen fouling in warmer weather. In a bedded system the dunging area may be at least as large as the lying allocation, but would be at least wide enough to permit convenient house cleaning by tractor.
- The merits of varying feeding methods are debated in Chapter 5. However it is important to provide sufficient space, 200 mm (8 in) for restricted feeding and 50 mm (2 in) for ad-lib feeding. Hopper design is also critical; a 100 mm (4 in) high by 150 mm (6 in) deep feed space being essential for pigs in this category.
- For pigs in the grower stage the step between the lying and the dunging area should be no greater than 100 mm (4 in). Suitable slatting materials

are concrete up to 100 mm (4 in) slats with 17 mm (¾ in) gaps, mild steel, woven wire, expanded metal and plastic; probably in that order.

Although the layouts discussed are commonly used, attention is drawn to the opening section of this chapter which details the important operational advantages of regular movement of pigs into rooms suited to their size and environmental needs.

LAYOUT 30

Bedded Kennels—Tractor Cleaned under Monopitch Roof

Important Considerations

The kennel size shown would be adequate for up to twenty-eight pigs at 20 kg but only half that number at 40 kg.

The dung passage should be the minimum width to allow for a commercial tractor: headroom with safety cab needed may be 2.8 m (9 ft 6 in).

Note that a 150 mm × 100 mm (6 in × 4 in) high wide kerb is used over which the dung passage gate closes to protect the gate from damage when tractor cleaning.

The kennel would be 750 mm–900 mm (2 ft 6 in–3 ft) high and insulated.

The pophole dimensions would be 300–450 mm (1 ft–1 ft 6 in) wide × 450 mm–600 mm (1 ft 6 in–2 ft) high. Plastic flaps may be used in the winter to retain the heat within the kennel.

It is normal to position the water supply lines inside the kennel to prevent freezing in the winter, with the drinker positioned on the outer wall of the kennel and sited so as not to be fouled by the dung passage gate when closed.

The dung passage gates should be either tubular positioned at 75 mm (3 in) centres, or 50 mm × 50 mm (2 in × 2 in) heavy gauge welded mesh.

A full width × 900 mm (3 ft) deep feed flap would also be used for ventilation and it may be preferable to be able to remove the lids completely during hot weather to prevent fouling within the kennels.

Operating the System

The pigs are penned in groups of twenty-eight per pen initially entering the pen through the doorway into the centre passage.

The number of pigs per pen is reduced to the size compatible with the eventual number in the finishing house which may mean some remixing.

Hopper feeding is best suited to a bedded layout and this would be sited within the kennel area.

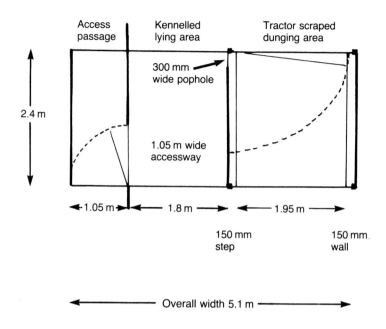

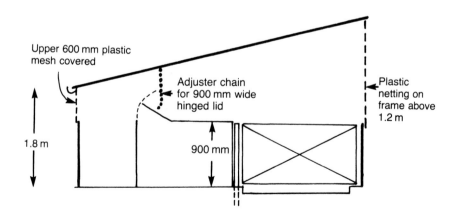

LAYOUT 30 Grower penning – bedded kennels with tractor cleaning under monopitch roof

Daily dung scraping is recommended to maintain a higher level of pen cleanliness with bedding added only to the lying kennel.

Merits of the System

The use of bedding acts as a good temperature buffer.

It can be fitted under a non-specialist structure.

Disadvantages

It is necessary to enclose the pigs regularly whilst cleaning and handling the bedding.

The building must be high enough at the eaves to allow tractor access for cleaning.

A fixed pen size demands some movement of pigs and/or floor space wastage.

Stock inspection is not easy.

Scraped passages may increase the spread of enteric diseases.

Layout 31

Tractor-cleaned Verandah

Important Considerations

This layout is as described in Chapter 10 (Layout 29), with the exceptions that:

1. The dunging area is 50 per cent larger to allow access for tractor cleaning.
2. The house is under a covered roof (insulated over the kennelled area) with space boarding above 900 mm (3 ft) on the outer walls which can be hinged down to enhance warm weather ventilation. This space boarded area is bottom hinged and 600 mm (2 ft) high.

Operating the System

As described in Layout 30.

Merits of the System

The use of a low insulated kennel helps to maintain the temperature required by the growing pig, particularly where plastic pophole flaps are used in cold weather.

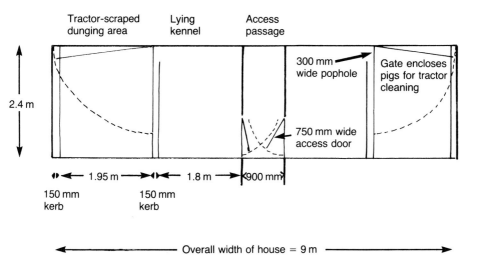

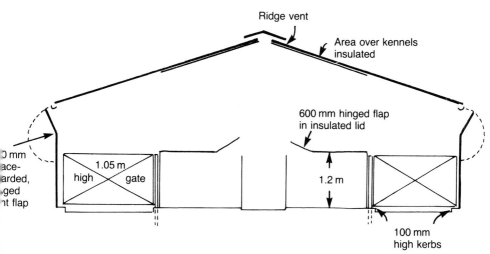

LAYOUT 31   Grower penning – tractor-cleaned verandah

Disadvantages

Piglet inspection is not easy.

Subdivision of groups poses management problems.

A specialist, small tractor or motorised scraper is required.

Cost Comparisons

These are summarised in table 11.1.

Table 11.1   Comparative costs of layouts described in chapter 11

Layout	Description	Narrative	Relative cost per grower housed
30	Bedded kennels with tractor cleaning	Includes hopper cost	110
31	Verandah part-slatted	Floor fed	100

Chapter 12

FINISHING PIG HOUSES

The Principles

THERE ARE a considerable number of criteria which need to be satisfied in selecting and appraising a finishing house.

A good starting point is to decide when the finishing stage is to commence, and then to assume that it will be completed at least 90 kg liveweight, or the heaviest probable future slaughter weight. To define these parameters at the outset is of considerable benefit as it determines the extremes of the spatial needs of the pigs and provides a greater opportunity for the provision of a more precise design of ventilation system and environmental control. Furthermore, it also helps to focus the problems of the finishing period: the wide age and weight range involved, the varying needs of the pigs and their vastly different tolerance levels of LCT and UCT. With a wide range of weights being involved the likelihood of operational inconveniences occurring is increased, such as the need to feed different diets to smaller pigs.

It is this rapid change in the size of the pig which creates the dilemma in housing design. This is also one of the primary reasons why so many finishing houses do not work to full expectation because too many compromises are made in space and climatic control and, possibly, in the stocking of such houses. The latter point is well illustrated by the very commonly observed example: it is possible to house twenty-five pigs at 40 kg in a pen that will hold eighteen pigs closer to slaughter weight. Furthermore, such is the growth potential of pigs that in relatively few days they are overstocked, growth efficiency is impaired and house operation becomes increasingly difficult.

There are various ways of reducing the problems caused by the rapid growth of pigs in this phase which include:

- Using two or three houses with even larger pens and moving the pigs as they grow.
- Subdividing houses into sections with each 'room' having pen sizes and a ventilation system appropriate to the weight of pigs in each section.
- Having several houses of the same basic size and design and using them on a batch basis to allow between-batch cleaning and the restocking with pigs of a closer weight range.

The real problem on a majority of farms is that the 'finishing house' is a single airspace house of a large number of pens of the same size with pigs sharing that house over a three-month weight range. Such a system means that the temperature, air change and space allowances are geared to the *average* size of pigs in the house which means that they are *wrong* for the majority of animals housed. In addition, it will be close to impossible ever to contemplate a hygiene programme because the house is so large that it cannot even be planned to depopulate it.

When trying to answer the primary question related to finishing house design which is 'what weight range?', the best advice is to ensure that the system chosen is at least capable of being subdivided, even if the pens are the same size. Then temperature control can be tailored to the pig sizes and numbers, and the appropriate attention given to the various classes of pigs—*and* between-batch hygiene becomes a possibility.

Another important consideration, already referred to in Chapter 2 is the question of deciding on the number of pigs in a house (or house section) or pen. Starting with a theoretical or ideal approach the answer may be 'as few as possible'. A more practical application is to say that a pen or section of a house would contain pigs of a weight range no greater than 10 kg from the largest to the smallest. With regard to the number of pigs in a pen, the same philosophy of the 'fewer the better' should be applied, with every effort made to keep the weight range as narrow as possible to maintain a more even growth rate and, thus, make more effective use of the pen space by marketing the pigs simultaneously.

There are other advantages of using smaller group sizes. All trials and experiments show that the fewer the pigs in a pen the faster and more efficiently they grow. This does not mean that every pen of pigs grouped in smaller numbers will grow faster than every larger group, but the average will be better and there is likely to be less trouble with bullying and vices. When considering batch size it may be appropriate *not* to exceed pigs born more than two weeks apart in the same air space, to subdivide those pigs weaned or purchased each week into at least five pens, with the overriding proviso that no pen should hold more than twenty pigs and that ten per pen is more likely to give better average results.

General Considerations

Space

Whilst it is necessary to suggest guidelines for floor space allowances it is important to indicate that the feeding system, group size, effectiveness of temperature control, pen shape and dung disposal system can all cause a considerable variance with regard to defining 'optimum' floor space allowance. The problem with defining the minimum allowance is that the speed of growth of pigs makes this figure rapidly inappropriate and it is only useful in judging compliance with recommended standards. It is important to recognise that space allowances do not just contribute to pig performance but also influence pig cleanliness. Both too much floor space (particularly with small

Table 12.1 Space requirements per pig for finishing pigs under varying conditions

Pig weight range (kg)	Lying area		Add feed space if not floor fed (assumes 300 mm (1 ft) deep feeder)		Dunging-area	
	Summer m^2 (sq ft)	Winter m^2 (sq ft)	Hopper m^2 (sq ft)	Trough m^2 (sq ft)	Bedded m^2 (sq ft)	Slatted m^2 (sq ft)
40–50	0.3 (3.2)	0.25 (2.7)	0.02 (0.25)	0.07 (0.75)	0.25 (2.7)	0.12 (1.3)
50–60	0.35 (3.8)	0.3 (3.2)	0.02 (0.25)	0.07 (0.75)	0.3 (3.2)	0.12 (1.3)
60–70	0.4 (4.3)	0.35 (3.8)	0.03 (0.32)	0.075 (0.8)	0.35 (3.2)	0.12 (1.3)
70–80	0.45 (4.8)	0.4 (4.3)	0.034 (0.36)	0.08 (0.89)	0.35 (3.2)	0.13 (1.4)
80–90	0.48 (5.2)	0.45 (4.8)	0.037 (0.4)	0.08 (0.89)	0.35 (3.2)	0.15 (1.6)
90–100	0.51 (5.5)	0.48 (5.2)	0.037 (0.4)	0.09 (1.00)	0.35 (3.2)	0.17 (1.8)

pigs in colder weather) and too little floor space (with larger pigs in warmer weather) will increase the fouling of the lying area.

Dunging area allocation will also depend upon:

Bedded passages. There is a minimum width through which conventional tractors can conveniently pass—2.025 m (6 ft 9 in)—and unless hand or automatic passage scrapers are used, this width may determine the dunging area allowance per pen.

Part-slatted pens. These rarely allow a clean lying area floor if they are less than 1.2 m (4 ft) deep and the full width of the pen.

Totally slatted pens. These are discussed more fully later in the chapter. They may permit a satisfactory reduction in the total floor area allocated due to little problem of pig cleanliness, but by no more than 20 per cent of the *total* shown for part-slatted penning in table 12.1.

Pen Shape
This may also influence floor space allocation in that certain principles of pen shape enhance pen cleanliness. The following points should be considered in deciding upon pen shape:

Feeding system. A greater scope in choice of pen shape exists with hopper or floor feeding systems, provided there is a minimum of 1.5 m (5 ft) between the hopper feed face and the opposite pen division. Because pigs must stand at right angles to a trough in order to feed there should be at least 1.8 m (6 ft) for pigs over 80 kg—and this dictates, to some extent, the overall shape of the pen.

Dung removal system. Where dunging areas are allocated across the narrower width of the pen the method of removal can determine the shape as well as the size of the pen. Long and narrow pens tend to create behavioural problems because pigs may find it more difficult to reach dunging and drinking points, which leads to fouling of the lying area and reduced water, and thus, feed, intakes. In general it is wise not to use pens which *are longer than twice the pen width*. Then it is important to apply the pigs per pen and space per pig recommendations made previously in this chapter.

Temperature

The need to control air velocity passing over the pig to prevent discomfort in cold weather has been stressed in Chapter 3. In addition, the effect of feed scale, group size and the use of bedding upon the temperature range for pigs must be taken into account. The lying area temperature figures for finishing pigs of 21°C up to 60 kg and 18°C for pigs over this weight must be adjusted to the other aspects of housing mentioned. No compromise on insulation or environment control in general can be tolerated in a finishing house system where pigs are kept in small groups, fed to a strictly rationed scale and without bedding in a house with inadequate draught-proofing.

Pen Divisions

Although lower pen divisions can be used where a trough is sited against a pen surround, the height of pens cannot safely be less than 1.2 m (4 ft) for pigs up to 100 kg or 1.05 m (3 ft 6 in) for pigs up to 65 kg. Where horizontal tubular divisions or pen fronts are used the gaps between the rails under 900 mm (3 ft) from the floor (or trough), should be no greater than 125 mm (5 in), but this can be extended to 200 mm (8 in) above that height. A 40 mm (1½ in) diameter tube is normally of adequate size for spans up to 3 m (10 ft). High tensile wire has also been used for this purpose but a higher maintenance need makes it unsuitable for pig housing.

Passage Widths

A passage used for access to pens for manual feeding only should be no less than 900 mm (3 ft) wide, though 1.05 m (3 ft 6 in) is preferable. If the passageway is to be used for driving pigs as well, then 1.05 m (3 ft 6 in), is the minimum, though 1.2 m (4 ft) would be preferable. It is important that the pen doors open to create a barrier across the passage for easier pig movement without the need to use a movable hurdle, and that such doors are hung to allow movement to the end of the house to and from which pig access is most likely to occur.

Tractor-cleaned Systems

A minimum passage width of 2.025 m (6 ft 9 in) has already been stated. In addition concrete kerbs of 150 mm × 150 mm (6 in × 6 in) are recommended to protect the walls and troughs and to allow the gates to close over during cleaning out. Such kerbs between the lying and sleeping area also help to prevent pen wetting and, thus, help keep the lying area clean. The use of gates in tractor-cleaned systems does raise potential problems of pig security in that smaller pigs may become trapped if the gap under a gate or between bars exceeds 100 mm (4 in). So in some tractor-scraped layouts it may be necessary to have a system whereby the gate is lower in its open position than when closed over a kerb for cleaning out. It should also be remembered that a medium-sized tractor with safety cab requires a height clearance of 2.8 m (9 ft 3 in).

Part or Totally Slatted Systems
The choice of slat for a relatively large number of pigs is narrowed to those which are robust and have a potential long life. Materials which fit this description include concrete and cast iron/mild steel as preferred materials. Concrete slats of 100 mm (4 in) width with an 18 mm (¾ in) gap between them should be specified or 50 mm (2 in) wide metal bars with 12 mm (½ in) gaps if metal is preferred.

Selecting the Layout

In Chapter 11 the principles of choosing between subdivisions of groups and their movement to larger pens has been discussed and, as for growers, the advantages of moving finishing pigs to larger pens with suitably respecified ventilation must be stressed.

Because the permutations of dung removal and feeding systems are so numerous it is neither possible, nor necessary, to show all the layout possibilities. The use of bedding has its advantages as discussed in Chapters 2 and 3, but not every producer is able to use it and handle it, so the same degree of individual choice due to circumstances and preferences would be applied to feeding systems. It is, however, relevant to draw attention to the increasing trend towards the adoption of totally slatted systems for pigs. It would be wrong to assume that this system is gaining in popularity because of convenience and simplicity of daily operational routine alone. The genetic improvement which has reduced backfat levels on modern pigs allows the possibility of ad-libitum feeding to be considered up to slaughter and this lends itself well to total slatting. Indeed pigs on hopper feeding systems remain much cleaner if pens *are* totally slatted. Further, disenchantment with fan ventilation and its capability to encourage pigs to dung in the lying area of a part-slatted pen has accelerated the trend to total slatting. In fact it is not easy to define circumstances where an unbedded, part-slatted system could be recommended for fattening except in naturally ventilated verandah or monopitch layouts.

Thus, the foregoing comments in this book and this chapter suggest that a preferred finishing house layout would fulfil the following conditions:

- It would be subdivided so that pigs of a weight range no more than ± 5 kg from the average are housed in the same air space.
- It would have preferably ten but no more than twenty housed in a pen.
- The house would be either bedded or totally slatted, or kennelled if part-slatted.
- Floor feeding would not be practised.

It should be noted that, except where specified, the stocking rates assumed in the following sections of this chapter relate to pigs at 90 kg and table 12.1 should be used for different weights of pig.

STRAW BEDDED SYSTEMS

Reference to the variations depending upon the feeding systems used is made in the notes and description of each system. Where versatility in the choice of feeding system is possible, this is also indicated.

The width of tractor-scraped passages shown in this type of system assumes that a conventional farm tractor is used. Narrower dunging passageways can be used if hand-steered motorised scrapers or a skid-steer loader are used. There are a few important considerations concerning tractor-scraped passages:

- If they are made narrower to suit a specialist cleaning machine there is a danger that pigs will be dirtier in warm weather when they may choose to reverse their lying pattern. The ratio of dunging to lying area should be as close to 1:1 as can be arranged and no less than 0.5:1.
- The length of passage that can be easily scraped with a tractor is around 45 m (150 ft) but it may be halved if a hand-steered scraper is used, or at least will demand that several passes down the passage are needed to clear the muck on each cleaning occasion.

For finishing pigs a 300 mm (1 ft) diameter trough can be used, or 450 mm (1 ft 6 in) if a double-sided arrangement is to be used. Feed hoppers should have a 125 mm (5 in) high feed face with each feeder section 185 mm (7½ in) wide and a 200 mm (8 in) deep feed face for pigs up to 100 kg.

Drinkers should be sited between 750 mm (2 ft 6 in) and 1.05 m (3 ft 6 in) from the tip of the drinker to the floor level depending upon weight of pig and drinker type. In naturally ventilated systems it is normal to run the water supply inside the kennel to prevent freezing and to use galvanised droppers for drinkers in the dunging area.

It is helpful in keeping passages drier that where bite-type drinkers are used they be inserted in a 'box' so the pigs can only use the drinker when standing directly in front of it. This prevents the drinker being operated from the side which creates greater spillage.

The flat-topped kennels shown in Layouts 32–34 are in extensive general use. However, recent experience suggests that kennels with a pitched roof line may ventilate more effectively, allowing better pig observation and easier addition of bedding. Such kennel tops do largely preclude straw storage.

Plate 15 Suffolk type finishing layout with tractor scraped yard and a bedded kennel area in which, in this case, pigs are trough fed.

Layout 32

Suffolk Kennels with Trough Feeding (Twelve Pigs per Pen)

Important Considerations

The outer rows of kennels have insulated, weather-sealed walls and insulated lids. The kennels may be constructed with a flat top to accept straw storage.

The main barn-type structure is uninsulated.

The length of the trough determines the number of pigs per pen; in this case twelve pigs for a 3.6 m (12 ft) gross trough length and consequently the lying area shown is more generous than in some of the other layouts.

It is possible to replace the trough with a hopper and handgate into the access passageway.

Operating the System

The pigs are placed in the pens via the dunging passage.

Feeding is done twice daily either by hand or pipeline (wet feeding).

The pigs are enclosed in the kennels for cleaning, probably on three occasions per week.

Weighing is carried out either in the dunging area, centre passage or specialist area outside the house.

The flaps on the kennel fronts are hand adjusted to give some temperature control.

Merits of the System
It is a simple layout with flexibility suitable for various feeding methods.

Disadvantages

Access to pens for weighing, removal, etc. is not easy unless the hopper and handgate replace the trough.

The inspection of the kennels is not easy.

If the kennels are constructed too large, ventilation becomes very difficult in cold weather and may encourage respiratory problems.

It is a relatively costly use of floor space due to large dunging area:lying area ratio (1.16:1), but this will help in keeping the kennels cleaner in warm conditions.

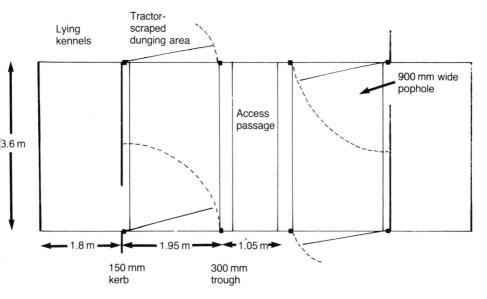

Lying
kennels

Tractor-
scraped
dunging area

3.6 m

Access
passage

900 mm wide
pophole

← 1.8 m → ← 1.95 m → ← 1.05 m →

150 mm
kerb

300 mm
trough

Overall width of layout = 9.45 m

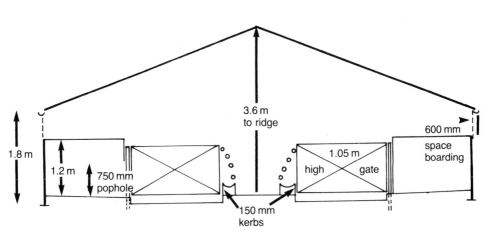

3.6 m
to ridge

600 mm
space
boarding

1.8 m

1.05 m
high   gate

1.2 m   750 mm
         pophole

150 mm
kerbs

LAYOUT 32   Bedded finishing penning – Suffolk layout with kennels,
            tractor-scraped (12 pigs per pen)

LAYOUT 33

Zigzag Kennel Layout (Twelve Pigs per Pen)

Important Considerations

For the best ventilation results the length of the kennels should not be more than twice the width, and if they are sealed at the rear, they should be ventilated naturally.

If mechanical (pipeline) feeding is used the outer passage could be made narrower as it is only used as an accessway.

It is possible to replace the trough with a hopper and handgate into the feed passages.

Outer building cladding should be solid up to 1.2 m (4 ft) above passage level, and space boarding or similar should be used above.

Operating the System

See Layout 32.

Merits of the System

The siting of the lying area in the centre of the house increases pig comfort and warmth.

Sited under a barn cover, the central kennels increase straw storage capacity.

Inspection into the kennels from the outer passage is better than in the Suffolk layout even if plastic strips are used for winter conditions.

Disadvantages

Unless a hopper and handgate to the access passage are used pig movement is difficult.

It has a slightly worse ratio than the previous layout of dunging:lying (1.22:1), and a second passage means that less of the house area is occupied by the pigs.

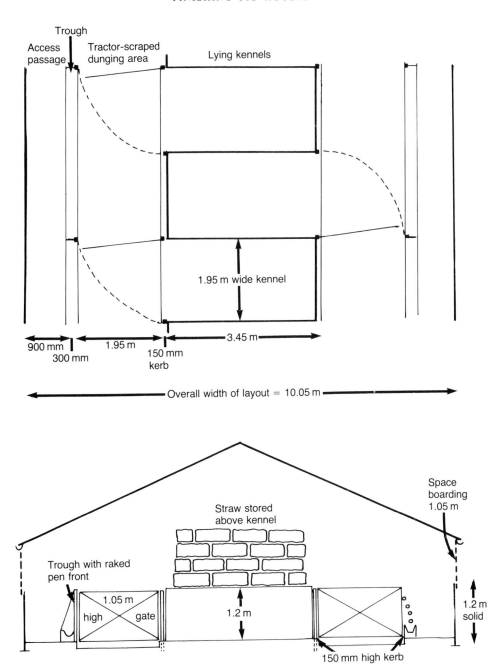

LAYOUT 33 Bedded finishing penning – zigzag kennelled layout with tractor-scraped passage (12 pigs per pen)

Layout 34

Kennelled 'Ulster' (Twelve Pigs per Pen)

Important Considerations

It is vital that the kennel length does not exceed twice its width. Whenever longer kennels are used ventilation control and respiratory conditions become a problem.

The dunging passage outer walls should be solid up to 1.2 m (4 ft) above the passage floor, and space boarding or similar up to the eaves.

Operating the System

As for previous layouts, but cleaning out may have to be conducted more often due to the smaller area per pen.

It is often designed with a handgate into the centre accessway for better pig handling.

It demands the automatic dispensing of feed (usually by pipeline).

Merits of the System

Better observation of the pigs is possible within the kennels via a 1.35 m (4 ft 6 in) hinged flap against the centre passage.

A single access passage layout and reduced dunging area: lying area ratio (0.58:1) gives a larger lying area for pigs.

Kennels constructed to the centre of the building improve pig comfort.

Disadvantages

It only works satisfactorily if the kennel length: width ratio is maintained as described.

A relatively narrow pen can lead to some fouling of the kennel due to the difficulty of pigs reaching the dunging area over prone pen-mates.

It demands auto-delivery of the feed and is not suited to hopper feeding.

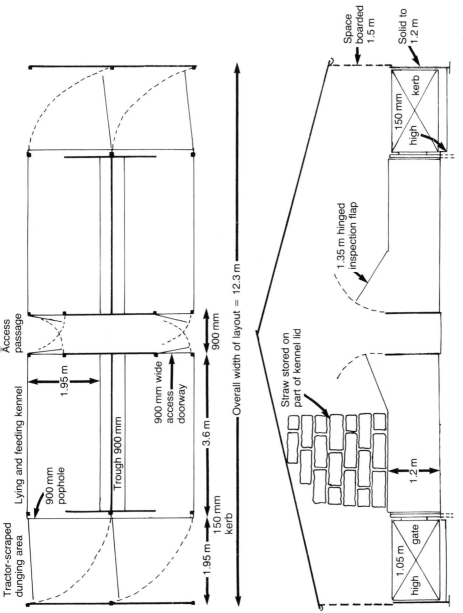

LAYOUT 34 Bedded penning – kennelled 'Ulster' layout with tractor-scraped passages (12 pigs per pen)

Layout 35

Central Dunging (Twelve Pigs per Pen)

Important Considerations

This layout may be used either with a kennelled lying area or within an insulated structure with mechanical or automated natural (ACNV) ventilation.

The pen shape is geared to the provision of a suitable trough allowance although the trough could be replaced by a hopper and handgate, although no hopper feeding works particularly well for finishing pigs in a bedded lying area due to the amount of straw which becomes pushed into the feed face.

Operating the System

The pigs are placed into the pen via the dunging passage unless a handgate to the access passageway is provided.

All other operations are as for the previous layouts except that a reduced dunging area will necessitate cleaning out more often than in the Suffolk or Zigzag kennel layout.

Merits of the System

It has a good dunging:lying ratio (0.58:1). There is good pig visibility in the lying area and it is one of the few bedded layouts where the dunging area of each pen can also be seen from the access passage.

Disadvantages

Although a smaller dunging area increases pig occupancy of a given area, this can increase the risk of lying area fouling in warmer weather.

Two feed/access passages are required.

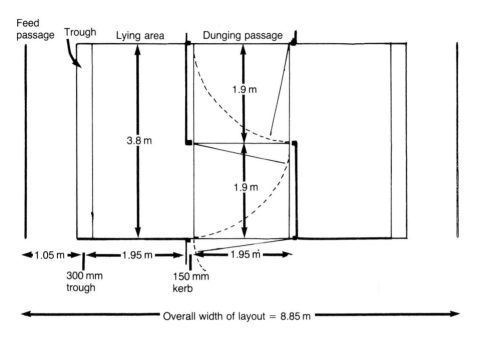

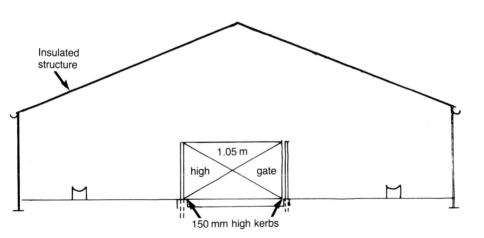

LAYOUT 35 Bedded finishing penning with central scraped zigzag dunging passage within insulated building (12 pigs per pen)

Layout 36

Deep-strawed Dunging (Twelve Pigs per Pen)

Important Considerations

The fall of the dunging area floor should be exaggerated to assist in the run-off of liquids to a shallow gully outside the yard over which a tractor loader can easily pass.

The ramped area may be replaced by a step and it is normal to allow up to 750 mm (2 ft 6 in) of muck to build up.

The lying area is normally kennelled with a hinged access flap against the feed passage and a handgate for ease of pig movement.

Feeding is either by hopper, as shown, or trough along the 3.6 m (12 ft) kennel wall. The hopper may be used to form the kennel front and is filled by an overhead conveyor.

A water point may be sited over a plinth with a drain and a pipe to the gully outside the pen gate.

Above the pen front gate it may be preferable to incorporate either space boarded flaps or a plastic mesh covered frame up to 3 m (10 ft) in height which is advisable for fore-end loader operation.

Operating the System

The pigs are placed in the yards via the handgate into the access passage if fitted, otherwise from the dunging area.

Muck is removed either between batches or as necessary. If the pigs are still in the pen when cleaning takes place, they should be enclosed in the lying area by use of a hurdle.

Bedding is added as required.

This system lends itself well to larger group/pen sizes.

Merits of the System

Daily routine is reduced as manure is only removed occasionally.

The first finishing system described without scrape-through dunging therefore allows better hygiene routines (between-batch cleaning) to be practised.

A larger dunging area may help to keep the kennel clean.

Big straw bales can be used to reduce handwork.

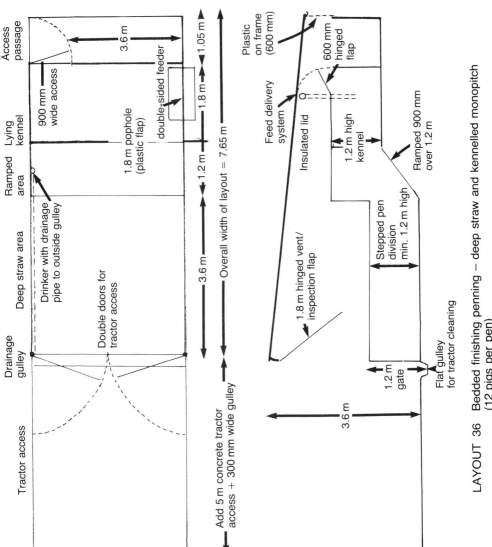

LAYOUT 36   Bedded finishing penning – deep straw and kennelled monopitch (12 pigs per pen)

Disadvantages

A relatively large area per pig is required to maintain cleanliness with infrequent cleaning out.

A large external concrete or hard area is needed for fore-end loader access. This is better if it is shared with a second house opposite.

There can be a problem with freezing water if it is sited near the main pen gate to reduce wetness.

It can be difficult to remove pigs from the pen for weighing or loading.

Pigs will wallow in this system in warm weather and, although this is natural, they will appear unacceptably dirty to many operators.

LAYOUT 37

Extended Monopitch (Twelve Pigs per Pen)

Important Considerations

This is similar to the Suffolk system shown in Layout 32, except that no cover building is needed so insulated, weather-proofed kennels are required.

The shape of the pen means that insufficient pen width is available for any other feeding system.

The kennels work best if they are provided with adjustable flaps above the 1.2 m (4 ft) pen front walls. The front height of the kennel is twice that of the rear.

Extra ventilation can be achieved by having a hinged or detachable section of the kennel top and an adjustable area between the uninsulated cover of the centre dunging/access area and the top of the monopitch lying area.

Operating the System

The pigs are placed in and removed from the pens via the handgate into the centre passage.

Feeding is either by hand or by overhead conveyor.

Tractor cleaning is done at least three times weekly.

Ventilation adjustments are as described above and may be automated (ACNV) for extra refinement.

The use of a hopper as part of the pen front helps to reduce costs.

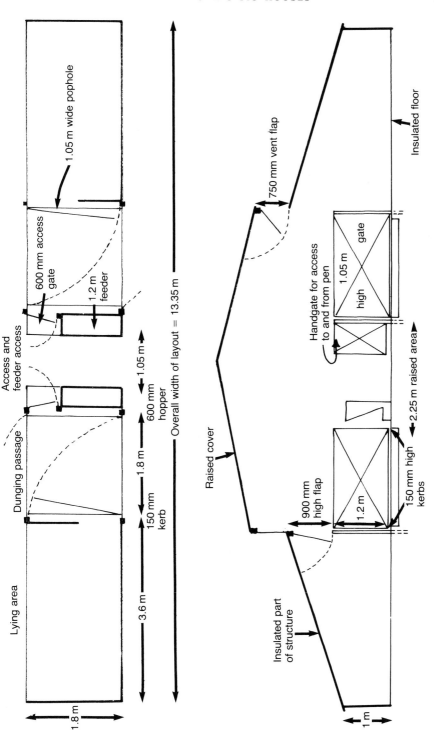

LAYOUT 37 Bedded finishing penning – extended monopitch lying areas with tractor-scraped passages and hopper feeding (12 pigs per pen)

Merits of the System

It has an excellent dunging:lying ratio (0.58:1) with a simple access passage.

It enables easy pig movement.

It has a simple construction.

Disadvantages

Pig observation may not be easy.

Hopper feeding demands pigs of good genetic capability or tolerance of higher backfat levels by the slaughter outlet.

PART-SLATTED SYSTEMS

The development of these systems had its origin in the need to increase output per man and in the improvement of factory built building construction which helped to reduce the costs of fully insulated structures, thus allowing smaller structures to be used for a given number of pigs.

All such systems run the risk of dirty lying areas if stocking rates and air movement are not precisely controlled, or the ratio of dunging area to lying area is too low. Floor feeding is widely used in these layouts, and is considered essential by many to achieve a manageable, clean lying area. The disadvantage of floor feeding in achieving an even growth rate and the tendency to overstock pens and, therefore, increasing vices have already been referred to.

A part-slatted layout which is well designed and operated intelligently can be low in daily labour needs, but if either of these provisos are not met, it may well reduce performance and be a potential additional source of criticism from those who advise that all pigs should receive some bedding.

It is unadvisable that any dunging area should be less than 1.2 m (4 ft) wide, and in naturally ventilated areas the ratio of dunging:lying is best kept close to 1:1, to avoid pen floor fouling. Drinkers should be sited over the slatted portion, or alternatively, low over the trough if that type of feeding system is incorporated.

Plate 16   This simple monopitch house is naturally ventilated and is shown with a partly slatted pen front. Hopper, trough or floor feeding may be used.

LAYOUT 38

Verandah Layout (Nine Pigs per Pen)

Important Considerations

An insulated lying area is required.

Ventilation slides which give an openable area of at least 0.42 m (4.5 sq ft) above pig level are required. Plastic flaps may be hung over the popholes to give better temperature control in the winter.

It is best sited at least 6 m (20 ft) from any other building to allow natural airflow.

Normally floor feeding is used, but hoppers may be used with auto-conveying but these would reduce the lying area available.

Operating the System

Pigs are placed in the pen from a central accessway which may also be used for weighing the pigs.

A low roof pitch makes the use of floor feeding dispensers somewhat difficult.

Slurry is removed as necessary according to the farm waste system installed.

Merits of the System

It is a simple construction with low running costs and a relatively low initial cost.

Good observation of the lying area is possible.

It is convenient for between-batch cleaning.

Disadvantages

If airflow is impeded by poor adjustment or surrounding obstacles floors tend to become dirty in warm weather.

Observation of the pigs is not easy if some are in the dunging area.

The removal of the pigs for weighing necessitates a long walk to the outside of each pen to enclose pigs in the lying area, and this is not easy for a single operator.

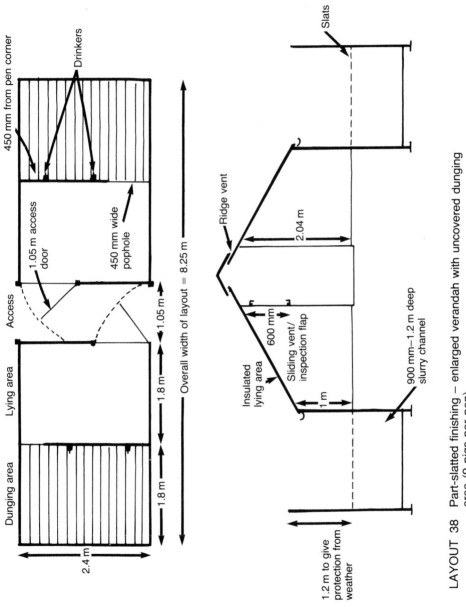

LAYOUT 38   Part-slatted finishing – enlarged verandah with uncovered dunging area (9 pigs per pen)

Layout 39

Naturally Ventilated Monopitch (Fourteen Pigs per Pen)

Important Considerations

The length of the pen must be no more than twice the pen width and front height twice that of the rear of the pen.

The front of the pen above the 1.2 m (4 ft) high door should have an adjustable flap which may be automated (ACNV).

Additional ventilation is via a hand-operated, hinged flap in the rear wall of the pen and through a hinged roof feed/inspection hatch which is normally 900 mm × 900 mm (3 ft × 3 ft).

Roof and walls of the building should be insulated as well as the floor.

It lends itself to floor, hopper and trough feeding. In the latter case the trough normally is sited along the entire length of the pen division.

It is best sited with the open front facing south.

Operating the System

The pigs are placed in the pens via the pen front which may open onto a race for ease of movement.

Hand feeding from an outdoor walkway is necessary if floor feeding is practised due to the low pen height. Auto-filling of hoppers is possible, and pipeline feeding with a trough system also works well with this layout.

Adjustments to flaps described above help to control the pen temperature and the pigs' dunging habits.

Merits of the System

It is a simple, low cost construction with few running costs.

Pig observation is excellent.

The completely divided pens allow between-batch hygiene to be conveniently practised.

It is very versatile as it can be used for pigs of very wide weight ranges. This layout may also be bedded and hand cleaned or totally slatted if either system suits the farm circumstances better.

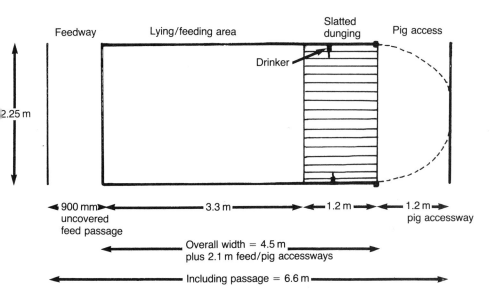

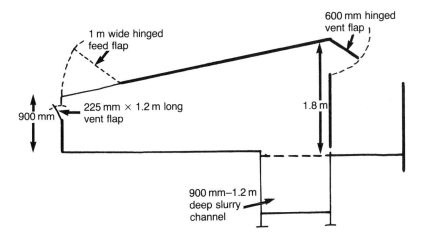

LAYOUT 39   Part-slatted finishing – naturally ventilated monopitch (14 pigs per pen)

Disadvantages

Inspection from the outside is unpleasant in bad weather.

It demands regular adjustment by the operator to control pig comfort.

The low ceiling height makes automatic floor feeding difficult to arrange.

Layout 40

Fan-ventilated 'Danish' Type (Twelve Pigs per Pen)

Important Considerations

It is normally fan ventilated with one of the systems described in Chapter 3, and the success of the system depends upon the accuracy of environment control.

The simple pen and house configuration is suited to floor or trough feeding. If trough feeding is used, alternate pen divisions may be replaced by a trough with a zigzag trough divider over it to allow two pens to use a single 450 mm (1 ft 6 in) wide trough.

A fully insulated structure is required.

Operating the System

The pigs are admitted to and removed from the pen via the handgate into the central passageway (which may also be used for weighing).

Feeding is at predetermined intervals either by hand or automatically.

Slurry is removed as appropriate to the farm disposal system.

Merits of the System

The simple pen layout keeps capital costs down.

It gives excellent pig observation.

It lends itself well to an automated feeding system.

Disadvantages

It is very dependent upon the design of the ventilation system and its good operation to maintain pen cleanliness.

Unless the house is sub-divided it is not easy to move the pigs for cleaning. A compromise on temperature level and air movement is then necessary.

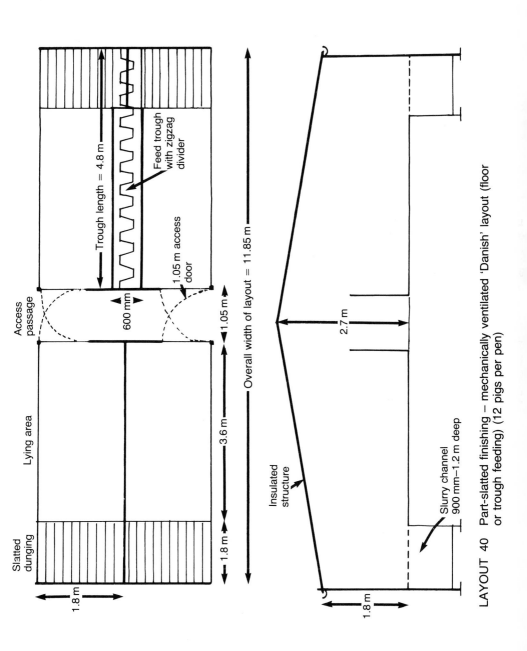

Trough length = 4.8 m

Feed trough with zigzag divider

1.05 m access door

600 mm

Access passage

Lying area

Slatted dunging

1.8 m

1.8 m

3.6 m

1.05 m

Overall width of layout = 11.85 m

Insulated structure

2.7 m

Slurry channel 900 mm–1.2 m deep

1.8 m

LAYOUT 40 Part-slatted finishing – mechanically ventilated 'Danish' layout (floor or trough feeding) (12 pigs per pen)

Layout 41

Central Zigzag Dunging Area (Twelve Pigs per Pen)

Important Considerations

A two-access-passage layout increases the area of house not occupied by pigs, but allows for a simple house layout.

A fully insulated structure is required.

Mechanical ventilation is needed.

It lends itself to any feeding system—floor, hopper or trough.

Operating the System

The pigs are admitted to and removed from the pen via the handgate into the access passage.

All other operations are as for Layout 40, but dung removal would be required at more frequent intervals due to the reduced capacity of the slurry chamber.

Merits of the System

Cheaper construction is possible due to a single slurry chamber.

It allows good pig observation.

Feeding can be easily automated.

Disadvantages

As for Layout 40.

If a trough is used in this layout the pig access to and from the passage is restricted.

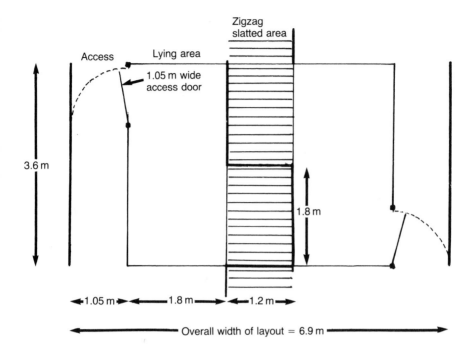

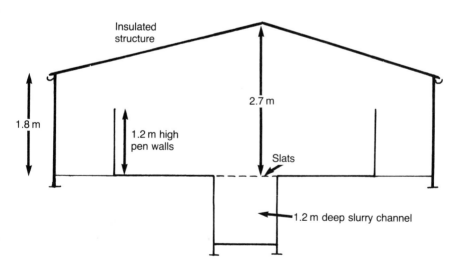

LAYOUT 41   Part-slatted finishing – with zigzag central slatted passage and fan ventilation (12 pigs per pen)

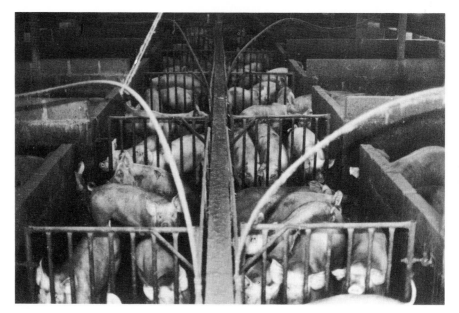

Plate 17   Part-slatted piggery with central slatted area between two lying areas where, in this case, pigs are floor fed.

LAYOUT 42

Ulster Trough-fed (Twelve Pigs per Pen)

Important Considerations

It is shown in the diagram as a single airspace, mechanically ventilated layout; but it may also be installed with a dividing wall between the lying and dunging area and as a naturally ventilated layout with hand adjusted slides.

A narrower central accessway is made possible because of the step-in walkways which allow feed to be placed in the troughs on either side and pigs to be inspected. Also it is possible to replace the step-in accessway by a mesh-covered raised walkway through which feed can be manually deposited into the troughs sited directly below.

Operating the System

The pigs are placed in and removed from the pens via the handgate into the central access passage. Low doors or bars may be needed to block the step-in accessway when pigs are being moved if this is not raised above access passage height.

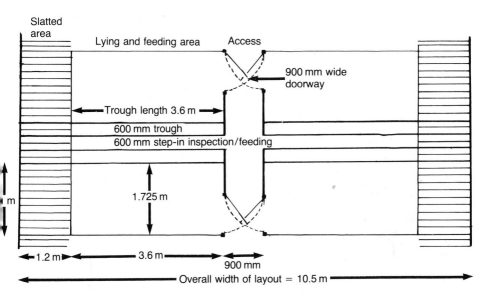

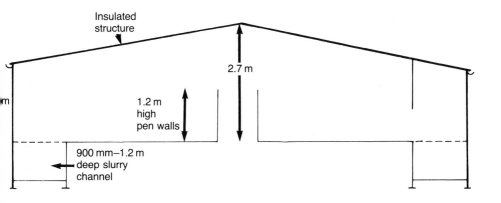

LAYOUT 42 Part-slatted finishing – trough-fed 'Ulster' layout with fan ventilation (12 pigs per pen)

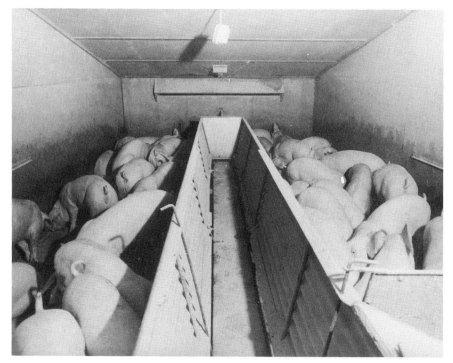

Plate 18   This part-slatted piggery has a step-in accessway to allow trough feeding on each side.

Manual or liquid (pipeline) feeding may be practised.

Slurry is removed as appropriate for the farm's disposal system.

Merits of the System

It gives the opportunity to trough feed without mechanical means and without extending the dunging area to a larger size by changing the pen shape.

It allows for excellent observation of the stock.

Disadvantages

As described above, there can be certain inconveniences in terms of pig movement.

Plate 19 Another example of a part-slatted layout with automatic floor feed dispensers and slats to the outer sides of the house.

TOTALLY SLATTED SYSTEMS

These have tended to become more widely adopted as disillusionment with floor cleanliness has followed inaccurate control of mechanical ventilation, and genetic improvements have given rise to more ad-lib feeding (which tends to lead to pigs fouling their lying areas). Total slatting does not mean that less good ventilation control can be tolerated. Indeed, as shown in Chapter 3, on totally slatted floors even better temperature control is required. However, it does remove the need to control *air direction* and so ACNV and less sophisticated fan control can be used.

Although the greater cleanliness of pigs is possible than with other systems, the temptation to overstock must be avoided. However, in practice, it is possible to reduce the overall floor area per pig, but if floor space allowances are excessively reduced the incidence of vices can be expected to increase.

Total slatting requires that careful consideration be given to the design of the slurry chamber otherwise separation of the solid fraction may occur. The regular subdivision of the below-slat area is recommended to ensure good flow of the slurry together with the use of 'honeycomb' support walls for slat support and pen divisions.

Layout 43

Paired-pen Layout (Eighteen Pigs per Pen)

Important Considerations

It is often designed with fan extraction over the centre accessway, and with some manual control over the air inlet openings along the eaves wall of the house.

The centre passage is often also slatted to remove the need to clean after weighing and/or pig movement. Depth of chamber beneath the centre accessway is often deeper than that under the pigs and is used to accept slurry from the pens via a sluice gate.

A double-sided hopper is easier to use if filled by an overhead conveyor.

A fully insulated structure is required.

Operating the System

The pigs are handled via the centre accessway.

Feeding and slurry removal are convenient.

It may be used in conjunction with a subdivided section with smaller pens, so pigs remain within the same structure from weaning through to slaughter.

A pair of pens have the ventilation adjusted to the size of the pigs in each section.

Merits of the System

It gives excellent pig observation from the step-in walkway or stable-type door.

The subdivision of the house allows differential temperature control and between-batch hygiene to be practised.

Disadvantages

Total slatting adds to the initial cost of the system.

The odour levels may make operation less acceptable.

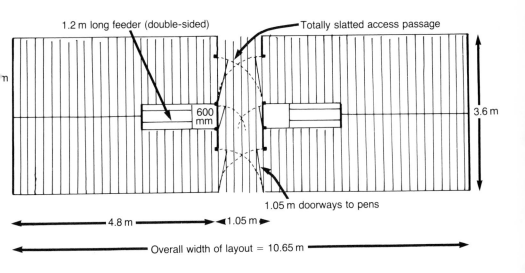

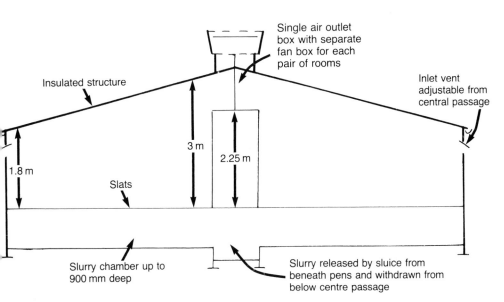

LAYOUT 43   Totally slatted finishing – paired pens on a 'room' principle (18 pigs per pen)

LAYOUT 44

Trough-fed Layout (Sixteen Pigs per Pen)

Important Considerations

A 450 mm (1 ft 6 in) double trough with zigzag trough divider is used between a pair of pens.

Other layout considerations are as for the previous layout.

It may be divided into rooms of four pens, i.e. one pair on each side of central passageway, or multiples of four pens.

Operating the System

Pig access is via the central passageway.

If the house is subdivided, between-batch hygiene can be practised.

Slurry is normally removed from the side walls with a slurry channel divided below each pair of pens.

Merits of the System

There are fewer divisions than in Layout 43, but it will only work as well if it is subdivided to allow differential setting of ventilation/temperature control.

A denser stocking rate is possible than in the part-slatted variant of this layout (Layout 40).

Pig observation is excellent.

Disadvantages

The pigs have to be brought in and removed from the house via rooms occupied by other pigs.

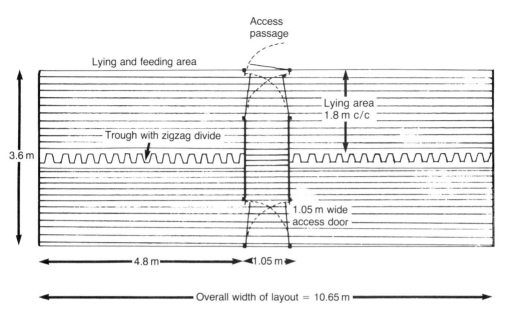

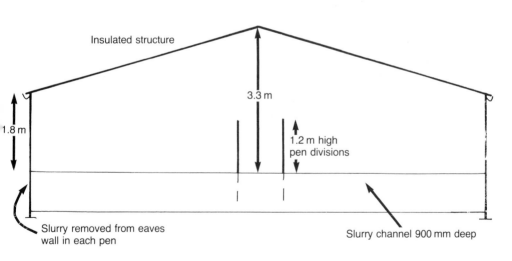

LAYOUT 44   Totally slatted finishing – trough-fed, fan-ventilated layout which may be subdivided into rooms (16 pigs per pen)

Layout 45

Room Principle House (Twelve Pigs per Pen)

Important Considerations

The room sizes are geared to the herd size. In the example shown with four pens × twelve pigs per pen this is suited to the weekly output of around 120 sows weaned at approximately three weeks of age.

Each room has floor space suited to the upper average weight of the pigs to be housed. The pigs are moved on every two to four weeks depending upon the frequency chosen. It is feasible to plan a house to incorporate a weaner section and then to move into the layout shown at around 25 kg; the pigs are then moved to sections with progressively larger pens and refined ventilation as they grow. Clearly use of this principle demands precise calculation prior to construction.

There is a choice of hopper or trough feeding (as in Layout 41).

Operating the System

The pigs are moved into smaller penned rooms at 25 kg and then moved up the house approximately every three weeks.

The pigs are moved along the slatted centre accessway.

Slurry is removed from the eaves wall to allow subdivision of the house between rooms.

Merits of the System

The tailoring of pen size to a narrower weight range ensures a more efficient use of floor space without rebatching or mixing the pigs.

Differential temperature and ventilation settings can be used to suit the pigs in each section.

Between-batch hygiene can be practised.

Excellent observation of the stock is possible.

Feeding can be easily automated on dry or wet principles.

Disadvantages

Although remixing of the pigs is almost eliminated they do need regular movement.

It is necessary to market slaughter pigs over a three-week period so that rooms become vacant for pig movement.

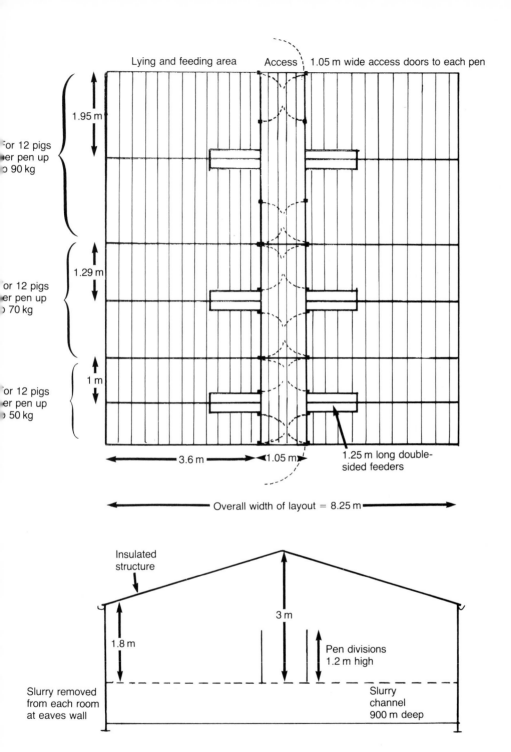

Lying and feeding area   Access   1.05 m wide access doors to each pen

1.95 m

For 12 pigs per pen up to 90 kg

1.29 m

For 12 pigs per pen up to 70 kg

1 m

For 12 pigs per pen up to 50 kg

3.6 m   1.05 m

1.25 m long double-sided feeders

Overall width of layout = 8.25 m

Insulated structure

3 m

1.8 m

Pen divisions 1.2 m high

Slurry removed from each room at eaves wall

Slurry channel 900 m deep

LAYOUT 45   Totally slatted finishing – house divided into rooms with larger pens to suit pigs as they grow (12 pigs per pen)

Plate 20   Totally slatted finishing pens with feeding via hopper in this case.

Cost Comparisons

These are summarised in Table 12.2.

Table 12.2   Comparative cost per pig place of layouts described in chapter 12

Layout no.	Description	Variation	Relative cost
32	Suffolk with trough feeding		150
32a	Suffolk with trough feeding	Automated feeding	175
33	Zigzag kennel layout	Trough, manually fed	140
33a	Zigzag kennel layout	Automated feeding	148
33b	Zigzag kennel layout	Auto-filled hopper	148
34	Kennelled Ulster with auto-trough feeding		145
35	Central dunging with trough		140
35a	Central dunging with trough	Auto-feeding	165
36	Deep straw with hopper		140
36a	Deep straw with trough	Auto-feeding	165
37	Extended monopitch and hopper	Auto-filled	130
38	Part-slatted verandah—floor fed		100
38a	Part-slatted verandah-hoppers	Auto-filled	108
39	Monopitch—part-slatted	Hopper-fed	105
39a	Monopitch—part-slatted	Trough—auto-fed	125
40	Part-slatted Danish—floor fed		135
40a	Part-slatted Danish—floor fed	Auto-fed—trough	160
41	Central part-slatted dunging—floor fed		120
41a	Central part-slatted dunging	Auto-fed—trough	145
42	Ulster, part-slatted—trough fed		135
43	Totally slatted paired layout	Hopper fed	150
44	Totally slatted with central trough	Auto-fed	180
45	Totally slatted 'room' layout	Hopper fed	160
45a	Totally slatted 'room' layout	Auto-fed—trough	185

Note: not all these variations are shown in the layout diagrams.

Chapter 13

OPERATIONAL CONSIDERATIONS

LOADING RAMPS

UNIT SECURITY and ease of operation both assist in the long-term efficiency of a unit. It is important to give thought to the ease with which pigs can be loaded to leave the unit as well as reducing the potential contamination risk from a lorry. The peripheral siting of a loading ramp has already been mentioned (see Chapter 2), and races which make one-man movement of pigs possible are also recommended. It is advisable to have pigs turn through 90 degrees prior to entering the crush or holding area at the foot of a loading ramp, so that they are less likely to turn round and return towards the operator. It is also advisable to have a narrow loading pen with at least two gates so that pigs can be more easily restrained. In a fixed-ramp system the tailboard rest should be 500 mm (1 ft 8 in) below the loading ramp floor and should be 900 mm (3 ft) long and slope away from the unit. This detail will help to prevent debris from the lorry entering the unit. The ramp should have a gentle incline of some 1:4 or 1:5.

Getting pigs onto the upper deck of lorries may be made easier by:

- A winch-operated ramp.
- A hydraulic lorry tailgate lift.
- A variable height loading platform onto which pigs are driven and which is operated by a tractor hydraulic system.

WEIGHING AND HANDLING

There are advantages in being able to weigh and handle pigs easily so that the tasks are conducted with minimum effort and to best effect. The occasional weighing of batches of pigs allows a check on growth efficiency to be made and changes to management to be monitored. In addition it allows weight tolerances in slaughter pigs to be met without incurring price penalties.

As described in the various layouts, many pigs will be weighed in the house accessways. However, many systems would benefit from a separate area, particularly where a large, bulk weigher can be used to allow batches of pigs

Plate 21   A good weighing layout allowing pigs to be funnelled to the scales.

to be checked. Electronic devices can be adopted so that weights are easily registered on a digital scale with a printer facility as an additional option.

Important Considerations

- Races for pig movement should be no wider than a pig requires to turn round in, and smooth sided so that an operator can move pigs single-

Plate 22   The pneumatically operated gates on this weigher has a print out and automatic pig marking device to speed weighing.

handed using a movement board. The advised width is 1 m–1.05 m (3 ft 3 in–3 ft 6 in), with divisions 1.2 m (4 ft) high. Races regularly used for cull sows may need to be slightly wider—1.2 m (4 ft).

- A fixed weigh area should have either a narrow feeder pen or one with an adjustable gate to permit one-man operation or to improve work effectiveness, by allowing one man to weigh and mark pigs whilst the second returns the weighed pigs to their pens and fetches a pen still to be weighed (see Layouts 46 and 47).

- Large weighers have the advantage of permitting groups of pigs to be checkweighed and, alternatively, allowing individuals for slaughter to be weighed separately and allow room for the operator to enter the scale to work and slapmark the pig before releasing it.

- It is also feasible to site the weigh area so that it is not only joined to the piggeries via races but also to the despatch/loading point.

In Layout 46 two weight layouts are shown. Layout A is suited to a larger herd and would facilitate some check weighing of groups of pigs. This is a distinct management aid in a finishing enterprise. The two holding pens allow one operator to actually weigh the pigs and record the data whilst a second returns pigs to their pen and fetches a further group to the vacant holding pen.

Layout B lends itself to single-operator use with a conventional portable pig weigher.

The combined weighing and handling area shown in Layout 47 is again intended for single-operator use.

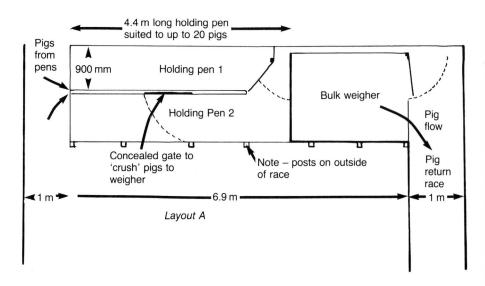

4.4 m long holding pen
suited to up to 20 pigs

Pigs from pens

900 mm

Holding pen 1

Holding Pen 2

Bulk weigher

Pig flow

Concealed gate to
'crush' pigs to
weigher

Note – posts on outside
of race

Pig return race

1 m

6.9 m

1 m

Layout A

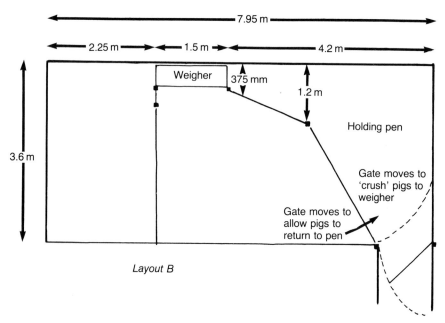

7.95 m

2.25 m

1.5 m

4.2 m

Weigher

375 mm

1.2 m

Holding pen

3.6 m

Gate moves to
'crush' pigs to
weigher

Gate moves to
allow pigs to
return to pen

Layout B

LAYOUT 46   Two pig weighing areas

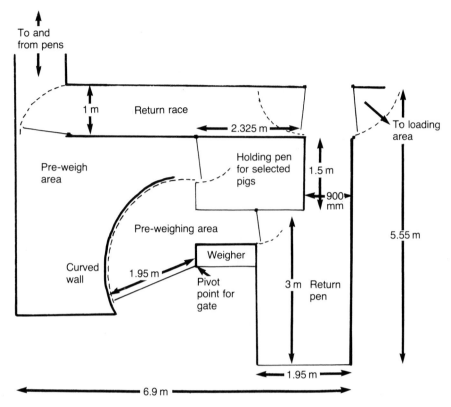

LAYOUT 47   Weighing and handling area (third option)

Carcass Disposal

Another unit requirement is a means of disposal of dead pigs—an unfortunate but inevitable consequence of pig production. There are two main methods for the larger unit:

1. *Incinerators*. These are normally propane fired and can be a most hygienic form of disposal of small piglets up to 10 kg. However, this weight limitation and the smells that are emitted make them a less than complete solution and they have a high running cost.

2. *Dead-pit*. This is a deep hole either block-lined or constructed of large diameter sewer pipes with an earthen base and airtight lid. This can be used where there is a low water table and no risk of contamination is likely. It relies on microbiological breakdown of the pit contents. Provided that the pit is not overloaded (two used in six-monthly cycles are best), the carcasses decompose completely, with little nuisance, over a few months.

ACCLIMATISATION PENNING

The need to protect the resident herd from the risk of infection by newly purchased pigs means that acclimatisation penning should be provided. Husbandry techniques to 'vaccinate' the new arrivals via dung or debris and contact with selected animals from the main herd are advisable.

Such penning should be sited peripherally to allow reasonable separation from the main herd and, ideally, would possess a separate drainage system to reduce contamination risks. It is important that such penning should be as secure as the main unit itself.

Ideally, the pigs will be fed, bedded and inspected from outside the pen so that any disease which might be incubating within the pigs is not transmitted to the resident herd by the operator.

Layout 48 incorporates pens into which all pigs or those close to slaughter weight are placed. The nose-to-nose contact allows the newly acquired stock to receive contact and so build up immunity levels to certain organisms prior to movement to the main herd. This layout is a straightforward bedded kennel layout which would be cleaned out at the end of the acclimatisation period, which may typically be twenty-eight days.

The number and size of pens will depend upon the herd size. As a guide, for every hundred sows it will be necessary to purchase forty replacement gilts and three boars per annum if all gilts are bought-in. Thus, the acclimatisation area should be large enough to hold (40 ÷ 12) approximately four pigs, *plus* resident pigs for every hundred sows if replacements are acquired monthly.

It is important to remember that because the pigs are stressed when changing farms and because small numbers are involved, good temperature control is essential in this type of housing and this is greatly helped by generous use of straw bedding.

BULK FEED BINS

The bulk storage of feed has become widespread due to increased herd sizes, even where manual feeding is used. The need to have trouble-free storage is critical in order to maintain diets in a contaminant-free state.

The first important feature of a good bulk bin (see Layout 49) is that it contains a dispersal cone or plate to prevent the blown feed sticking to a particular part of the bin wall. In addition there should be a breather pipe to allow excess dust to be collected and to reduce risks of condensation. There should also be an inspection hatch complete with a ladder with safety rings to allow checks to be made on the bin and to facilitate the six-monthly cleaning of the bin.

The hopper should itself be smooth sided and free from internal pro-tuberances which may disturb the flow of the feed. The low cone area should slope with sides of at least 45°.

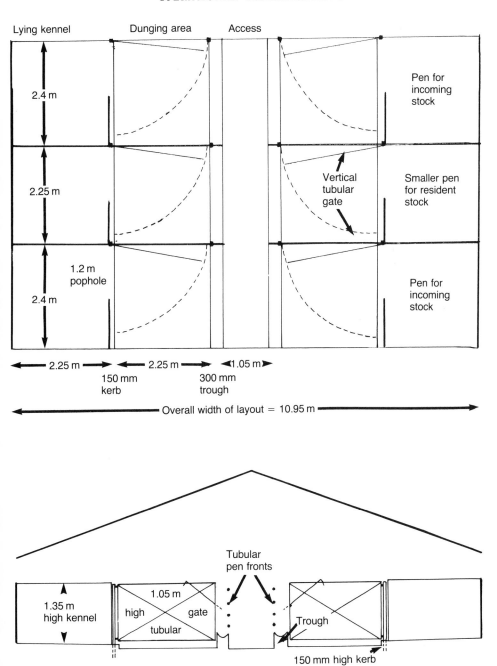

LAYOUT 48   Acclimatisation penning on bedded principle under a covered building
(maximum 10 pigs per pen)

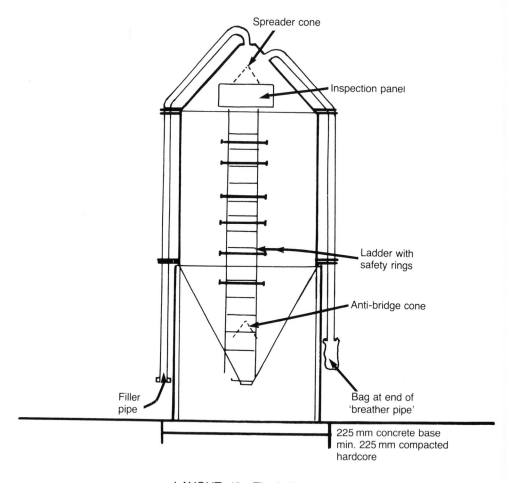

Spreader cone

Inspection panel

Ladder with
safety rings

Anti-bridge cone

Filler
pipe

Bag at end of
'breather pipe'

225 mm concrete base
min. 225 mm compacted
hardcore

LAYOUT 49   The bulk bin

STAFF FACILITIES

The importance of maintaining satisfactory staff performance means that
adequate facilities should be provided for their comfort. It is most desirable
that consideration be given to the provision of suitable office facilities so that
unit recording and monitoring is efficiently conducted. Furthermore, the
control of drug usage and cleanliness of equipment requires that an
appropriate storage and cleaning area which can be locked by senior
personnel is also included (see Layout 50).

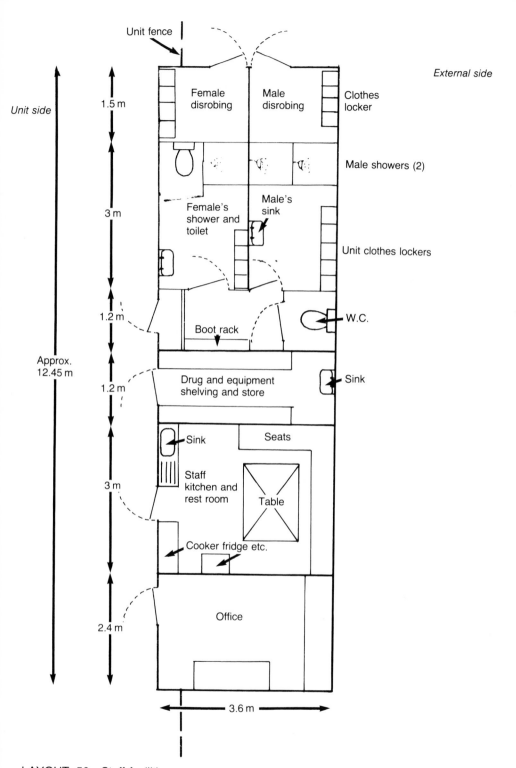

LAYOUT 50 Staff facilities

For reasons of unit health security as well as social comfort, there is an increasing need to provide shower and changing facilities. These should be so laid out that there is *no* opportunity for staff or visitors to enter the unit other than by the shower/bathing point. This approach to security also implies that staff will remain on-site throughout the working day, so it is necessary to provide a galley area in which food can be stored, prepared and eaten.

Added refinements to such a layout would include equipment for the washing of unit clothing, sterilising of incoming goods (in a formaldehyde gas chamber) and a workshop for repairs and maintenance on the unit.

MONITORING BUILDING PERFORMANCE

The relative success of a building will depend almost as much upon the method of operation employed as the design. Where problems of pig output or comfort arise there is little hope of a permanent solution being found to operation or structural modification if the manner in which the house is used is not closely monitored. It will be necessary to know the stocking rates, controller settings, prevailing weather conditions, etc. to effect a lasting solution.

Simple additions to pig records or diary notes may be needed for a period to enable an accurate picture of operational matters to be built up to achieve such a log of data.

PLANNING MAINTENANCE

Although it is rather too glib to suggest that pig producers expect their pig buildings to withstand the inevitable battering from pigs without the need to ever replace components or repair walls, it is realistic to accept that wear and tear will arise.

The repair of deteriorating materials (rather than a sudden breakage) can be achieved much more satisfactorily if it is planned into the unit operation. There are examples of times when this kind of work can be programmed without creating excessive disruption to herd operation. The list below is by no means exhaustive but the examples given show how opportunities to make space available for routine repairs can be taken:

- During warm weather use buildings normally occupied by other stock to hold pigs temporarily.
- Turn gilts or groups of sows into paddocks in mild weather during the day.
- Plan replacement or additional penning so that it is available slightly before it is required so that existing buildings may be repaired.
- Take advantage of favourable marketing circumstances to sell extra pigs to free penning for maintenance.

It is somewhat daunting to plan maintenance at a time when unit development is being contemplated. However, the capital-intensive pig business demands that not only is an appropriate design and construction method chosen, but that, once constructed, every effort is made to extend the effective life of that investment.

Chapter 14

PIG HOUSING AND LEGISLATION

EACH COUNTRY has its own approach to legislation covering the siting and erection of buildings, and the degree of legislation imposed is also subject to varying interpretation by local authorities.

In the United Kingdom two kinds of control are exercised. The first deals with the development of the land itself and this makes buildings subject to planning approval from the local planning authority. The second area concerns the actual structures used and their design which are administered by the Building Control Department of the local authority who should be consulted to determine what, if any, parts of the proposed works require approval.

Planning Approval

In the United Kingdom guidelines for the use of land is covered by the *General Development Order of 1977 (Class VI)*.

Whilst there are potential dangers in paraphrasing an Act of Parliament the main features of the Order are first summarised and then shown in full. It is most advisable that any building project is at first outlined to the local planning authority to seek their guidance and opinion as to whether approval is required.

Firstly the building proposed must be subject to classification as 'requisite' for the use of the land. Some authorities interpret this such that an intensive pig unit on a large farm is exempt from planning approval, whereas a smallholding does not require the incorporation of such an enterprise. Others require that *any* pig building requires planning approval and produce guidelines for farmers.

Given a satisfactory interpretation of that initial point, the current legislation exempts certain buildings from the need to have planning approval. Again it is important, at the development stage, to remember that legislation is subject to change, interpretation will vary and that some areas, for example Scotland, have different Acts. Currently, however, if the holding is greater than 0.4 hectares (1 acre), planning permission is *not normally required* for:

1. A building or exterior with a ground area of no more than 465 m² (5,000 sq ft) which is erected within 90 m (300 ft) of the nearest building providing more than two years have elapsed since the previous development.
2. The height does not exceed 12 m (39 ft) unless it is near to the perimeter of an airfield when it may be 3 m (10 ft).
3. The building is no closer than 25 m (82 ft) to a classified road.

There is also an important distinction between a pig building and a services building or structure such as an above-ground slurry store or an earth bank. A bank would be classed as 'works' and would not be considered as part of the area covered by buildings. It is important to note the insertion of the time clause, because if a building of 300 m² had been erected only one year earlier, 465 m²–300 m² = 165 m² could be added within the two-year period without being subject to planning approval.

It should also be noted that certain structures designed for another purpose moved onto a pig unit for use by pigs or as a store, do require planning approval because they were not designed for agricultural purposes initially.

Even if the above exemptions apply, the location of a Listed Building may prohibit building if the local authority considers such development would impose unacceptable restriction on the listed structure. Areas of National Parks and outstanding natural beauty or scientific interest also have special restrictions which supersede these exemptions.

Thus, planning seeks to safeguard the use of the land, and to control the development and the appearance of the countryside.

To avoid misinterpretation, the General Development Order (which is itself a section of the Town and Country Planning Act 1971), is given in full below. It is worth noting that the development of bare land for pigs may still be exempted from the need for application for planning permission provided that the use of such land is considered 'requisite' or essential to the farming business. The precise wording of the schedule for completeness is:

Class VI   Agricultural buildings, works and uses.
1. The carrying out on agricultural land having an area of more than one acre and comprised in an agricultural unit of building or engineering operations requisite for the use of the land for the purposes of agriculture (other than the placing on land of structures not designed for those purposes or the provision and alteration of dwellings), as long as:
(a) the ground area covered by a building erected pursuant to this permission does not, either by itself or after the addition thereto of the ground area covered by an existing building or buildings (other than a dwelling-house) within the same unit erected or in course of erection within the preceding two years and wholly or partly within 90 metres of the nearest part of the said building, exceed 465 square metres;
(b) the height of any building or works does not exceed 3 metres in the case of a building or works within 3 kilometres of the perimeter of an aerodrome, nor 12 metres in any other case;

(c) no part of any building (other than moveable structures) or works is within 25 metres of the metalled portion of a trunk or classified road.
2. The erection or construction and the maintenance, improvement or other alteration of roadside stands for milk churns, except where they would abut on any trunk or classified road.
3. The winning and working, on land held or occupied with land used for the purposes of agriculture, of any minerals reasonably required for the purposes of that use, including—
(i) the fertilisation of the land so used, and
(ii) the maintenance, improvement or alteration of buildings or works thereon which are occupied or used for the purposes aforesaid,
so long as no excavation is made within 25 metres of the metalled portion of a trunk or classified road.

It must also be stated that Article 3 (3) of the order qualifies the general permission given by the schedule in the following manner:

(3) The permission granted by this article and Schedule 1 to this order shall not, except in relation to development permitted by classes IX, XII or XIV in the said Schedule, authorise any development which requires or involves the formation, laying out or material widening of a means of access to an existing highway which is a trunk or classified road, or creates an obstruction to the view of persons using any highway used by vehicular traffic at or near any bend, corner, junction or intersection so as to be likely to cause danger to such persons.

Building Regulations

The second type of control is much more complex and deals with the building itself. This seeks to ensure that a building is structurally safe and complies to the Health and Safety at Work legislation, fire regulations and thermal insulation standards. The Code of Practice for *building regulations* is to be found in a comprehensive British Standards document number BS 5502. This was brought into effect in 1980 and it is hoped that this Code will be adopted by, or at least form the basis of, a standard for the whole of the EEC. Although British Standards do not have legal force as such, it is necessary that they be complied with to ensure that planning or building authorisation can be obtained and may be required in order that some structures qualify for any Grant Aid. Although standards for Buildings Regulations were laid down in 1976 in practice BS 5502 is now widely adopted for agricultural purposes.

BS 5502 is a weighty document and any new building should be considered against the guidelines of the Code itself. In truth there is some danger in providing a synopsis of the Code and, short of reproducing it in full, it is best that advice be sought prior to design approval. Many manufacturers of package deal buildings already have approval for their design under the Code leaving site considerations and services (i.e. water, electricity and gas), for example, as the major areas requiring the consent of local authorities. Thus, whilst the structure and fittings may comply with BS 5502 *plans must still be submitted* to the local authority to ensure compliance with these other regulations and this *includes the conversion* of existing structures. Furthermore, approval for both Planning and Building Regulations approval should be sought in plenty of time before the scheduled commencement of work.

The British Standards Code provides details for the selection and use of materials and, where appropriate, their support and application. Much of the Code itself cross refers to other British Standards, in particular in respect of materials and the DIY farmer/builder should seek the relevant Code or the manufacturer's guidance before selecting materials for construction. Thus, whilst not actually legislative, BS 5502 is the 'bible' by which the construction of buildings may be assessed.

In the Code, buildings are classified according to their design life and that, in turn, encompasses the density and duration of the occupation of the buildings by operators and the siting distance from a road or domestic dwelling. The categories of buildings are shown below, but, as far as specialist piggeries are concerned, only shelters for pigs kept out of doors are likely to fall into Class 4, and almost all piggeries will come into Class 2, or possibly 3. Briefly the classes are as follows:

Class 1 Buildings unrestricted as to purpose and location. Design life fifty years.

Class 2 Human occupancy should not normally* exceed six hours a day at a density not exceeding two persons per 50 m². No part of the building should be nearer than 10 m to a classified highway or human habitation not in the same ownership. Design life twenty years.

Class 3 Human occupancy should not normally exceed two hours at a density not exceeding one person per 50 m². No part of the building should be nearer than 20 m to a classified highway or human habitation not in the same ownership. Design life ten years.

Class 4 Human occupancy should not exceed one hour a day at a density not exceeding one person per 50 m². No part of the building should be nearer than 30 m to a classified highway or human habitation not in the same ownership. Design life two years.

It seems likely that pig buildings are manufactured to Class 2, but this can vary and the manufacturer should be consulted before building work is commenced.

A farrowing house, which on a large unit could have a greater degree of human occupancy, may in fact have to be a Class 1 building in order to comply with the above. If a Class 2 building currently meets the unit requirements, consideration should be made as to whether a change of use of this building or the density of human occupancy may be made in the future.

---

* The maximum occupancy to be expected at any time of the year for the purposes for which the building was designed. The maximum occupancy and/or density may be exceeded for up to eighteen days in any one year.

This could mean that a Class 1 building should be provided in the first instance.

It is probable that the following bodies will be interested in the classification of the building:

The Local Authority, Building control Department.
The Health and Safety Executive in Connection with the Health and Safety at Work etc. Act 1974.
The Ministry of Agriculture, Fisheries and Food.

If you apply for a grant on the structure as a building designed to Class 4 these are considered to be temporary and not eligible. Also on the claim form there is a declaration to be signed which includes a statement that *the items have been carried out in accordance with the requirements of the Health and Safety at Work, etc. Act 1974 and any other relevant provision of Law.*

Slurry compounds constructed of concrete panels or similar will need to comply with the relevant British Standard 5502. The standard for building earth bank compounds will be found in the Ministry of Agriculture booklet 2273 entitled *The Storage of Farm Manures and Slurries.*

The Code proceeds to define the materials suitable for buildings and modifications to suit each Class, but does not seek to exclude any materials not granted a British Standard specification. However, approvals will be more readily granted where materials of a known standard are used. In addition to references to materials (see Chapter 4) specific requirements concerning the structural strength of buildings to withstand wind and snow loadings is also given. Location does affect the wind or snow bearings and the guidelines given in BS 5502 should be studied in making choice of materials and the strength of building framework used.

Building Regulations also include various items of legislation of which the would-be builder should be aware. Although BS 5502 itself contains guidelines that:

Untreated liquid wastes should not be allowed to drain away, either by surface or surface drains, into any runway, watercourse, open ditch, stream, river or to a groundwater body but should be disposed of or treated in such a manner that no unacceptable pollution can occur. Local health or the appropriate authority should be consulted before manure storage or disposal facilities are established.

The laws to which the 'appropriate authorities' may refer are numerous and all seek to prevent damage to existing natural water courses or sewerage systems. A case of Good Agricultural Practice under the 1974 Control of Pollution Act is being prepared for use in the United Kingdom.

Permission to discharge overflows from slurry or effluent treatment systems is unlikely to be given where the Biological Oxygen Demand (BOD) (a measurement of the purity or impurity of the discharge) exceeds 20 mg/litre. In practice this rules out most piggery wastes under the Rivers (Prevention of Pollution) Acts of 1951 and 1961 and the Acts (Scotland) of 1951 and 1955.

Strict conditions also apply, defined in the 1963 Water Resources Act (and the 1972 Northern Ireland Water Acts) concerning the discharge of Wastes into underground strata. Local water authorities have powers to apply such conditions as deemed necessary to safeguard water in underground strata. A further act applies to underground discharges within three miles of a coastline.

Separate Acts of Parliament also cover the right of discharge of farm wastes into local authority sewerage systems in England/Wales and Scotland. This can be undertaken with the approval of the local council but will depend upon quantities to be released and degree of impurity as well as the local authorities' capability to deal with the material.

Buildings Regulations also include considerations under the Public Health (Recurring Nuisances) Act of 1969. This empowers an authority to serve a prohibitive notice upon methods of waste disposal, if infringement of the Act occurs. This has obvious implications for odour as well as spillage.

Legislation also applies to the discharge of smoke and fumes (Clean Air Acts of 1956 and 1968), but this is only likely to apply to pig producers who build chimneys for large incineration plants, swill/waste food boilers or who aim to provide heating by using straw burners or the like. The installation of all new furnaces must be notified to the local authority prior to commencement of works and the emission of smoke and fumes is subject to clear-cut regulations.

The security of electrical services is also governed by various Acts and there is an existing code of practice for all electrical installations. This code is continuously revised by the Institute of Electrical Engineers (IEE) and no electricity board will connect the mains supply to any installation which fails to comply with the current IEE regulations. BS 5502 specifies that every building should contain a diagram of the distribution circuits and control switches by the main supply switch. It should be added that rodent control is of vital importance both from the health standpoint and to safeguard electrical fittings. All electrical fittings should be subject to careful protection and choice of non-corrosive protective coverings to the environmental circumstances which prevail in piggeries.

Farmers are also subject to certain requirements under various statutory acts concerning safety at work. These range from the clear signing of dangerous or hazardous tasks concerning fragile roofs, fumes, floors and surfaces. Buildings must also comply to noise tolerances, toilet facilities and be adequately lit. There are specific requirements for strength, accessibility, dimensions and guarding of ladders, catwalks and walkways and, in particular, that any accessory 1.5 m (5 ft) above floor level must have a suitable handrail fitted. Ladders and roof ladders are also subject to clearly defined fixing and security statutes which include the directive that any ladder longer than 6 m must be fitted with safety hoops at not less than 900 mm (3 ft) intervals.

Further specific advice is given concerning the security of pits and slurry lagoons. These must be surrounded by a fence no less than 1.3 m (4 ft 6 in) high totally enclosed with approved woven wire mesh. Above-ground stores less than 2 m (6ft 6 in) in height must also be secured by suitable fencing;

slurry tanks should possess access ladders with a removable lower length, and warning signs to alert operators to the potential dangers.

When building plans are being laid it is worthwhile to allow the local Health and Safety Executive the opportunity to comment on proposals to secure the well-being of stock and operators.

All farmers are also required, under the Prevention of Damage by Pests Act 1949 and Rats and Mice Destruction Act of 1919, to maintain buildings, drains and sewers in a manner that reduces the likely infestation of the premises by rodents and this includes demolished or part-demolished buildings or siteworks.

Considerations of fire risk are also vital. Many materials used for the building structure, even though they are not combustible, may emit poisonous gases or drip when exposed to flame. The provision of escape panels which can be speedily removed from buildings to release stock in the event of fire are recommended, as is the provision of a static water point, or at least, the clear marking of ponds or other suitable sources for emergency use of fire fighting services.

As many buildings erected will be subject to a contract the prudent farmer will ensure that the work to be undertaken will be subject to the stipulations of these various acts and codes. This shifts some of the responsibility to the designers and manufacturers, but it will still be the farmer's duty to ensure that approval for the development of the site and the proper conduct of siteworks, services and materials used is given. The DIY builder/farmer should at least secure himself a copy of BS 5502 from the BSI Sales Department, 101 Pentonville Road, London N19 ND or seek advice from Ministry of Agriculture officials or Local Authority Buildings Office or professionally qualified persons.

Having made acquaintance with the legal requirements concerning buildings, it will be clear that there is no substitute for the early compilation of a thorough set of plans and a detailed specification in order that approval with regard to both Planning Application and Building Regulations are received. The local authority will regularly visit the site during works in progress and ensure that buildings regulations and local bye-laws are being adhered to.

Useful Reading and References

The following publications are used by the author in his own work and provide valuable reference for the interested reader who may require more detail.

Chapter 1

Intensive Pig Production, Baxter (1984), Granada Books, London.

Pig Housing, Sainsbury (1963—Revised), Farming Press, Ipswich.

The former of these two books, in particular, provides copious research references which form the basis of design detail.

MAFF Explanatory Leaflet, *Farm and Horticultural Development Scheme* (FADI).

All MAFF publications are available from Lion House, Willowburn Estate, Alnwick, Northumberland, NE66 2PF.

Welfare Code Recommendations (1983), MAFF Leaflet no. 702.

'Some Implications of the Welfare Codes for Designers and Users of Pig Buildings', Summers, *Farm Buildings Progress*, April 1984, pp. 11–13.

All Farm Buildings Digest and SFBIU references are available from Scottish Farm Buildings Investigation Unit, Craibstone, Aberdeen, AB2 9TR.

Farm Finance and Fiscal Policy (1976), Agricultural Mortgage Corporation.

Capital Investment and Layout Requirements, James (Nov. 74), NAC Conference Paper.

'Appraising Farm Buildings Investment', Crabtree and Pack, *Farm Buildings Progress*, Jan. 1983, pp. 13–16.

Chapter 2

'Farm Electric Centre Booklet' on electricity needs for mill/mix, slurry, etc.

Farm Buildings, Volume 1, Weller, Published by Crosby Lockwood.

MAFF 'Booklets on Slurry', nos. 2073, 2021, 2126, 2200 and 2356.

Planning Farm Buildings, Farm Buildings Information Centre Report no. 6. (NAC, Stoneleigh, Kenilworth, Warwicks.)

Colour Finishes for Farm Buildings, The Design Council, 28, Haymarket, London SW1Y 4SU.

Think Before You Build, MAFF leaflet no. 835.

'Calculating Sow and Weaner Accommodation', Robinson, *Farm Buildings Progress*, April 1981, pp. 19–20.

Farm Buildings—The Planning Approach, Cermak, SFBIU, March 1974.

How to Organize a Farm Buildings Contract—Wight, SFBIU, March 1975.

CHAPTER 3

MAFF 'booklets' nos. 2410 (Pig Environment), 2287 (Energy Conservation).
ACNV for Pig Housing—(1984) SFBIU.
'Design Temperatures for UK', Bruce, *Farm Buildings Progress*, Oct. 1983, pp. 5–7.
'Heating Requirements for Growing and Finishing Pigs', Bruce, *Farm Buildings Progress*, April 1984, pp. 15–19.
Natural Ventilation for Livestock Housing, Farm Buildings Information Centre Report no. 15 (1974).
'Keeping Pigs Cool', Fuller, *Pig International*, July 1980, pp. 6–14.
'Effect of Departure from LCT on Group Postural Behaviour in Pigs', Boon, *Animal Prod.*, 1981, pp. 71–79.
Handbook on Design of Ventilation System using Step Control and Automatic Vents (1977), Randall, NIAE Publication no. 28.
    All National Institute of Agricultural Engineering (NIAE) publications are available from Wrest Park, Silsoe, Bedford, MK45 4HB.
Influence of Environment on Utilization of Energy in the Pig, Close, Conference Paper.
'Temperature and Behaviour in Pigs', Health, *Behavioural and Neural Biology*, no. 28 (1980), pp. 193–202.
LCT in Control of Piggery Environment, NIAE Leaflet.
    The following Farm Electric Centre (FEC) leaflets are invaluable and available from FEC, National Agricultural Centre, Kenilworth, Stoneleigh, Warwicks.
Control of Vents in Livestocks Housing—FEC Tech. Info. Agri 7–4.
Essentials of Farm Lighting, FEC Handbook no. 25.
Installation Techniques for Lighting of Farm Buildings, FEC Tech. Info. 5–1.
General Lighting Layouts for Farm Buildings of Common Sizes, FEC Tech. Info. 5–2.
Daylighting of Farm Buildings, FEC Tech. Info. 5–6.

CHAPTER 4

'Large-Scale Fire Test on Insulation Materials for Farm Buildings', Kelly, *Farm Buildings Progress*, Jan. 1984, pp. 29–36.
Farm Buildings Cost Guide, Wight, SFBIU (Annual updated publication).
    Various Cement and Concrete Association Booklets available from Wexham Springs, Slough, Bucks., in particular:
Developments in Concrete Floors and Finishes (1978).
Farm Construction, no. 407–505.
Concrete Mixer for Farm Use—no. 7.
Insulated Floors for Piggeries—no. 3.
Repair of Concrete Floor Surfaces—no. 8.
Fixings to Concrete—no. 10.
Cost Effective Brickwork in Farm Buildings, Brick Development Association.

CHAPTERS 7–12

'Sow Service Accommodation', Robinson, *Farm Buildings Progress*, April 1981, pp. 21–24.
Blueprint for 200 Sow Unit—MAFF (1975).
'An Evaluation of the Farrowing Crate', Clough, *Farm Buildings Progress*, April 1984, pp. 21–26.
'Piglet Farrowing Boxes', Robertson and McCartney, *Farm Buildings Progress*, Jan. 1980, pp. 15–16.

'Piglet Foot Dimensions', Mitchell and Smith, *Farm Buildings Progress*, Jan. 1978, pp. 7–9.

Electric Underfloor Heating, FEC notes, April 1981.

Pig Finishing Houses, MAFF leaflet no. 46 (1971).

Also publications from the NAC Pig Demonstration Unit, Stoneleigh, Kenilworth, Warwicks.

CHAPTER 14

BS 5502 and Appendices, BSI Sales Department, 101 Pentonville Road, London. N1 9ND.

Health and Safety Hazards Associated with Pig Husbandry, H and S Executive Guidance note G530 (HMSO).

INDEX